How to Grow Psilocybin Mushrooms at Home

The Complete Step by Step Home Cultivation Guide of Magic Mushrooms.

JONATHAN SYRIAN

© **Copyright 2021 by Jonathan Syrian**
All rights reserved

This document is geared towards providing exact and reliable information with regards to the topic and issue covered. The publication is sold with the idea that the publisher is not required to render accounting, officially permitted, or otherwise, qualified services. If advice is necessary, legal or professional, a practiced individual in the profession should be ordered.

From a Declaration of Principles which was accepted and approved equally by a Committee of the American Bar Association and a Committee of Publishers and Associations. In no way is it legal to reproduce, duplicate, or transmit any part of this document in either electronic means or in printed format. Recording of this publication is strictly prohibited and any storage of this document is not allowed unless with written permission from the publisher. All rights reserved. The information provided herein is stated to be truthful and consistent, in that any liability, in terms of inattention or otherwise, by any usage or abuse of any policies, processes, or directions contained within is the solitary and utter responsibility of the recipient reader. Under no circumstances will any legal responsibility or blame be held against the publisher for any reparation, damages, or monetary loss due to the information herein, either directly or indirectly.

Respective authors own all copyrights not held by the publisher.

The information herein is offered for informational purposes solely, and is universal as so. The presentation of the information is without contract or any type of guarantee assurance.

The trademarks that are used are without any consent, and the publication of the trademark is without permission or backing by the trademark owner. All trademarks and brands within this book are for clarifying purposes only and are the owned by the owners themselves, not affiliated with this document

Table of Contents

Introduction to Mushrooms 1
    Mushroom Usage in Prehistoric Times... 2
    History of Mushroom Cultivation.. 3
    Types and Classification of Mushrooms 4
    History and Background of Psilocybin Mushrooms 5
    Further Improvement in the Discovery of Psilocybin Mushrooms..... 6
    Reproduction and Lifecycle of Mushrooms 7
    Spore Detachment 8
    The Spore Dispersal 9
    The life Cycle of Psilocybin Mushrooms... 10

Occurrence of Psilocybin Mushrooms 11
    How and Why Psilocybin Mushrooms Got Those Different Genes. 12
    Habitats of Psilocybin Mushrooms... 13
    Grasslands.. 14
    Dung Deposits 15
    Burned Lands 16
    Riparian Zones and Disturbed Habitats 16

The Biology of Psilocybin Mushrooms 19
    The Life Cycle of Magic Mushrooms.. 20
    Identification of Psilocybin Mushrooms... 21
    Why They Are Magic Mushrooms 26

Properties of Psilocybin mushrooms 30
    Chemical and Physical Properties of Psilocybin............................. 31
    Physical and Mental Effects of Psilocybin and Psilocin 32

Medicinal Properties of Psilocybin Mushrooms 35
    Biological Impacts... 36
    Hallucinogenic Effects of Psychedelic Mushrooms: 39
    Medical Uses 41
    Psychological Effects... 45
    How Psilocybin and Psilocybin Mushrooms Are Sold 49

Some Popular Psilocybin Mushrooms of the World 52
    The Genus Panaeolus.. 52

The Basic of Cultivation... 62
    Parts of a Mushroom.. 62
    The Lifespan of Theses Fungus: 65

Life Cycle of Psilocybin Mushroom 66
Reproduction of Mushrooms 68
The Basic Growing Parameters 70
Harvesting Hacks for Psilocybin Mushrooms 73
Making Spore Syringes 74
Agar Spore Germination 76
Cardboard Disk Spore Germination 77
PF-TEK 79
Flat Cake Tek 101
Rye Grain Tek 104
Popcorn Tek 107
Popcorn Tek Supplies and Equipment: 107
Making Grain Spawn Out of Popcorn 108
Fast Food of the Gods Method 113
Psilly Simon's Method 115
Tek for Magic Truffles - Truffle Tek 124
Cultivation Equipment and Supplies 137
Micro-Dosing Psilocybin Mushrooms 150
How Do You Micro-Dose With Psilocybin Mushrooms? 151
Step by step instructions to Take Your Micro-dose: 152
What Micro-dosing Schedule Should I Follow? 152
Advantages of Micro-dosing: 154
Leadership and Micro-dosing: 156

CHAPTER ONE

Introduction to Mushrooms

Mushrooms have been a part of the human diet for centuries. They have been a part of meals even before their proper discovery. People have been using them in their special meals and medicines, but they didn't know the actual name of it. Mushroom and its use as edibles by humans have a long history. With the development in food style, the style of using mushrooms was also changed. Humans are familiar with mushrooms long before the use of microorganisms. Mushroom is a fungus, and a proper definition of it was presented by Chang and Miles, where they said that a mushroom is a macro fungus which has a distinctive fruiting body, either epigeous or hypogenous, which is sizeable enough to be seen with the eye and picked by hand. They were not only used in Europe but also in the early Roman Empire, in Middle America, South America, and in Asia as part of special meals. Before the cultivation of mushrooms by man, they were found in the sediments of Lakes in Germany, Switzerland, and Austria. People didn't know anything about those umbrella-shaped puffballs found in moist places in prehistoric times. In ancient Greece and Rome, the toughest and hardest mushrooms like truffle and orange were the most expensive mushrooms.

Mushroom Usage in Prehistoric Times

Prehistoric times had evidence of mushroom usage in meals and medicines, but the exact time and place of its discovery are not clear. The Saharan aboriginal tribes of North Africa were possibly using mushrooms, as their stone painting showed, which were carved around 9000BC. In Spain, similar paintings were discovered, carved 6000 years ago. This evidence showed that some mushrooms were used in some specific rituals, and they were part of cultural meals as well. As per Egyptian culture and history of mushrooms, Egyptians considered mushrooms as "plants of immortality." Egyptian Pharaohs seemed interested in mushrooms so much that they announced and proclaimed that it is a special fruit meant only for gods. Therefore, in Egypt during the reign of Pharaohs, mushroom was considered an essential part of royalty meals.

Other civilizations of the world also considered mushrooms as something having superpowers. In Russia, China, Latin America, Mexico and Greece, mushroom was an essential part of some of their rituals. At some point, people thought that eating mushrooms could produce superhuman powers, and can lead a person on a path towards the gods.

History of Mushroom Cultivation

The cultivation of mushrooms was started back in 600s, in Asia. But in Europe, the fungus (mushroom) cultivation was started in the 1600s. History shows that the cultivation of mushrooms in Paris was started in the 1650s when a native farmer saw growing fungus in his field of melons. He then thought to cultivate this fungus on a commercial level, and this was the time to change thoughts about mushrooms. He delivered this umbrella-shaped fungus to restaurants where it was used as a dish. That type of mushrooms was named "Parisian Mushroom."

Researchers found fungus growing in caves around Paris, whose history dates back in the 1600s when French gardeners found that the humid and cool environment of caves is suitable for mushroom cultivation. Therefore mushrooms were cultivated at large scale in caves. In the Netherlands, mushrooms were introduced for cultivation in the 1800s but in small scales. Then after some time, large scale cultivation of mushrooms was done in marl mines in Limburg. After caves and mines, other methods of mushroom cultivation were developed, which then resulted in high mushroom yields in the region. Before that, mushrooms were specific, and they were only available for elites. Dutch people had the strictest rules and controls for mushroom cultivation. This was the reason the Netherlands became the largest mushroom producing country 50 years ago. However, China and the USA joined the competition of mushroom cultivation. China produces about 70% of the world's mushrooms now, so China is at the first position in mushroom cultivation, and then comes the USA and Netherlands.

Types and Classification of Mushrooms

Types of Mushrooms

In general, mushrooms are divided into four categories including, Edible mushrooms, Medicine or therapeutic mushrooms, Poisonous mushrooms, and Miscellaneous (mix) mushrooms.

Edible mushrooms: These mushrooms can be used in the human diet. They have been a part of meals for centuries in different civilizations. Some examples of these mushrooms are Cantharellus cibarius, Hericium erinaceus, Boletus edulis, etc.

Medicinal or Therapeutic mushrooms

Hundreds of mushroom species have therapeutic and healing properties, and therefore many of them are used in medicines. Some mushrooms having hallucinogenic properties are also considered as natural healers. For example, an old healer from Mexico, Maria Sabina, used hallucinogenic Psilocybin mushrooms to heal mental, spiritual, and physical problems. Examples of some therapeutic mushrooms are Fomitopsis pinicola, Hericium erinaceus, inonotus obliquus, etc.

Poisonous Mushrooms: They are extremely poisonous, which can even lead to death. In the next section, there is a detailed discussion about poisonous mushrooms. An example of such poisonous mushrooms is Amanita phalloides.

Miscellaneous mushrooms: There is a large number of mushrooms whose features and special characteristics are yet to be discovered.

History and Background of Psilocybin Mushrooms

As mentioned in history, some mushrooms have been used for religious purposes and in different rituals. These mushrooms, which have been of great importance according to a religious point of view, are mostly Psilocybin mushrooms. These mind-catching bodies have been an important part of cultures deprived of time and space. Many religions and philosophies of the world have been influenced by these "Magic Mushrooms."

Reko, Richard E. Schultes, Roger Heim, and R. Gordon Wasson are the lucky scientists and researchers who are responsible to re-discover the remains of ancient usage of Psilocybin mushrooms. They discovered that the old native Mesoamerican people used Psilocybin mushrooms in their rituals and ceremonies. Their research showed that modern-day mushroom cultures have ancestral links with the Psilocybin mushrooms, which were used in the religious practices of some civilizations like Aztec and Mayan. History shows that the Aztec people had so much belief in a specific Psilocybin mushroom

(Psilocybe Mexicana) that they named it Teonanacatl, which means "God's Flesh." A Franciscan in the 16th century went on an expedition with a team and reported about the beliefs of Aztec about "God's Flesh," but misguided Catholics. Catholics were running a campaign against Paganism at that time, and they were misguided about the usage of mushrooms. They restricted the use of Psilocybin mushrooms as they thought that pagans worshipped these mushrooms. This all resulted in the planned demise of this culture, but they couldn't destroy some evidence. The research and reports by Franciscan Bernardino De Sahagun are the ultimate sources of knowledge about usage of this type of mushrooms by Aztec tribes.

Further Improvement in the Discovery of Psilocybin Mushrooms

This situation gave rise to a new belief, with a mix of Christianity and mushroom rituals. R. Gordon Wasson and his team also discovered an indigenous healer named Maria Sabina, who practiced the use of Psilocybin mushrooms to heal many mental, physical, and spiritual problems. People said that she had specific "Magic Mushrooms." In search of these, the people of America came to know about the Psilocybin mushrooms of Mexico. When more people came to know about Maria Sabina, her popularity reached outside of America, and many people searched for her due to her magical mushrooms.

Not only in Mexico and America, but the traditional and religious use of Psilocybin mushrooms had also been found in ancient European culture. The philosophers Aristotle, Plato, Homer, and Sophocles were seen in history participating in the ceremonies in a different temple Demeter, in front of the goddess of agriculture. Strange types of ceremonies used to be held in those temples where people believed that they got changed after spending a night there. A conference about mushrooms was held in 1977 in Europe in which different psychoactive mushrooms of the world were discussed, and researchers got a new thing to research on.

Wasson and his team were the first Ethnomycologists, while Jonathan and his team proved themselves to be the second generation of Ethnomycologists. Human history has dramatic effects of Psilocybin mushrooms and many more yet to come. If we talk about the history of natural cultivation of Psilocybin mushrooms, a French mycologist Roger Heim was the pioneer who cultivated many Psilocybin species successfully, in the 1950s. He used the materials which were brought by Wasson and his team from their journey. Then the work started, and many Psilocybin mushroom species were discovered until the end of the 1970s, and still there are many mushrooms yet to be discovered.

Reproduction and Lifecycle of Mushrooms

Most of the fungi species have a sexual reproduction process. Sexual reproduction occurs when a new organism forms by the recombination of male and female genetic material. In scientific language, this genetic material is in the form of gametes. The genetic material or gametes of fungi are spores. The structure of a spore is well defined after the invention of the microscope. Otherwise, spores cannot be seen with the naked eye. According to microscopic observations, the spore is a cell protected by an outer wall, in a compact shape, and it is able to keep itself dormant for years. It makes itself active only when it finds a suitable environment for it. Most of the fungal species grow in a moist and humid environment, so when spore finds such an environment, they start the reproduction process. Take the example of Basidiomycetes whose spores are present on their basidia, which have a structure like baseball-bat, and they are present in the lining of gills (of mushrooms). These basidia are in a beautiful pattern arranged on the underside of the mushroom cap. This cap or pileus is attached to a stem of cylindrical shape, and mycologists call it "stipe."

Spore Detachment

Spores are of purple color when observed under a microscope. The outer end of basidium, which is horn-shaped and protrude outside, is called sterigma. It is an interesting fact that air under the cap of mushrooms, around the gills, is cooler as compared to the air on the upper side. It is because of the evaporation mechanism where the upper side faces sun rays, and the moist air on the underside becomes cool. During the evaporation process, the water around spores becomes condense when the air gets cool, this process creates a droplet at the place where the spore joins with its stem. Droplet gets growing until it loses its tension and can't grow anymore. As a result, the water in the droplet spreads over the spore's body, which forces the spore to move towards sterigma. When it touches sterigma, it faces an elastic reaction, thus returning with force: we can say it faces a catapulted action by sterigma. As a result, spore tends to fall, and the wind takes it away from its source with many of other spores.

The Spore Dispersal

The wind or air takes the spores away and throws them in fields, walls or anywhere else. When these spores find a suitable environment, they grow; otherwise, they remain dormant for years. Spores spread by wind are also displaced by animals and birds. For example, wind threw the psilocybin mushroom spore in a field, a goat came and ate the spores with grass, but it can't digest those spores as they have good armor. When the goat passes feces, those spores come out as such, and if they find a suitable environment there, they grow into a new psilocybin mushroom.

Fungi, when growing, contain a large number of hyphae. Hyphae are defined as the long filamentous cells, tubular-shaped, which get divided at the upper tips, creating a fork-like structure. A network of hyphae is called "Mycelium." Mycelium appears as white, hair-like growths on food or soil surface (called a substrate). Most of the fungi remain undifferentiated in the form of mycelium; only some of them grow into structures like mushrooms, puffballs, bracket fungi, etc. You may observe that the surface which is affected by fungus start to degrade. Actually, there are digestive enzymes secreted by mycelium, which degrade the substrate into organic molecules, and then mycelium absorbs those molecules as food.

The life Cycle of Psilocybin Mushrooms

Spores, when converting into mycelium colonies, eat the substrate and produce more spores. Those spores tend to look for spores from other colonies, for mating. It is because fungi want to spread diversity, and for this purpose, it forms different mating types. Whenever spores of one gender find suitable spores of another gender, they start mating. The cells of mycelium are monokaryotic and have a haploid nucleus (half genetic material). This slow, fuzzy material, when fusing with another half genetic material (haploid nucleus), it forms a cell containing two nuclei, this forming dikaryotic cell. This process is mostly specific to Basidiomycetes, where the nuclei remain separated even in a single cell. This phenomenon is different from other organisms because, in this process, two nuclei remain in a single cell until they combine inside the basidium. In short, the fungus intercourse starts very early and ends very late. The process keeps on going until this mycelium transforms into a fruiting body like mushrooms. Following is a simple illustration of a mushroom life cycle, which represents the life cycle of almost all genera, but actually, this illustration is of the life cycle of Psilocybin mushroom.

CHAPTER TWO

Occurrence of Psilocybin Mushrooms

Psilocybin mushrooms contain two compounds: psilocybin and psilocin. So, Psilocybin mushrooms are hallucinogenic mushrooms due to the presence of these two compounds. If we look at other mushrooms and their evolution, these compounds are not present in all mushrooms. It means that when these mushrooms evolved, they didn't have such hallucinogenic properties. If we look at the evolution of mushrooms, they were not like they are now. They evolved from Pseudomonas and then puffballs evolved and then morels, and now the latest umbrella shaped mushrooms. In the same way, different mushrooms adapted different properties from environment. Psilocybin mushrooms, which are also called Psychedelic mushrooms or magic mushrooms due to their magical affects on human mind, have these psychoactive compounds, specifically. A study was conducted by a researcher, Jason Slot, at The Ohio State University. He claimed that these

Psilocybin mushrooms have wide biological lineage, and different morphological appearance.

Environment has stressors as well as opportunities, and these two factors are also responsible for physical and genetic changes in organisms. Horizontal gene transfer is a phenomena which occurs in nature to develop changes at genetic level by transferring genetic material from other organisms. Slot believed that Psilocybin mushrooms are those which got hallucinogenic genes from such gene transfers. The examination of psychedelic mushrooms paralleled with the examination of other fungi (which are not hallucinogenic) showed some differences in genetic material. All hallucinogenic mushrooms (Psilocybin mushrooms) share a cluster of five genes which were not present in other fungi.

How and Why Psilocybin Mushrooms Got Those Different Genes

Psilocybin mushrooms have characteristics to create hallucinations in humans as well as they have ability to send human mind into an altered state. This is due to the presence of Psilocybin, which is present due to that cluster of Psilocybin producing genes. Study showed that those genes may have been transferred from some fungus-eating insects. As these mushrooms grow in an environment with insects, animal manure, rotten woods and other things like these, insects used to eat fungus, and as a result, protection was developed in those mushrooms by production of Psilocybin. It not only alters human mind but it also affects insect's brain. It suppresses a neurotransmitter in insects which lowers the appetite so they don't tend to eat these mushrooms. Eating Psilocybin mushrooms doesn't kill insects but it alters their brain and working capabilities. History shows that Psilocybin mushrooms have also been used as insect repellents even when the human didn't know about their exact uses. Many Psilocybin mushrooms don't share much genetic material except this cluster of five genes which shows that they come from different biological backgrounds. Sharing of this psilocybin producing cluster of genes shows that this horizontal gene transfer occurred for protective and evolutionary purpose, for the survival of these mushrooms. So, decoding of these genes can open new doors of research, to study these genes and find their impacts on human mind and body.

Habitats of Psilocybin Mushrooms

Psilocybin mushrooms grow all over the world but mostly they are found in fields and forests. They are saprophytes like other fungi, so they cannot make their food like plants. Instead, they usually grow on dead plant material and were restricted to narrow areas before humans found them. Some ecological and natural destructions like land sliding, floods, volcanoes and hurricanes created suitable habitats for Psilocybin mushrooms, by leaving the environmental effects. These natural processes are also responsible for traveling of Psilocybin mushrooms from one place to other. When humans started cutting trees, these mushrooms found their place on wooden chips and rotted woods etc. They adapted themselves to the human environment so much, that they started growing in gardens, near houses, in fields etc. It shows the relation of Psilocybin mushrooms with human activities. Human activities and climate changes turned many green areas into deserts, but the growth of Psilocybin mushrooms continued, and they continued to adapt themselves according to environment.

Depending upon the evidences, Psilocybin mushrooms have five major habitats, including:

- Grasslands
- Dung deposits
- Burned lands
- Riparian zones (flooded areas)
- Moss lands

Grasslands

Grassland habitats, which are mostly wet and swampy lands, support the growth of Psilocybin mushrooms. These moist environments are very supportive habitats for fungi, therefore some Psilocybin mushrooms are specific to grasslands. In such habitats, mostly conic shaped, tall, thin and small Psilocybes grow, for example *P.strictipes, P.liniformans, P.semilanceata, P.mexicana, and P.samuiensis.* Most of them grow on matted grass bases, camel grass or on lemon grass. Some tryptamine producing grasses are also a good habitat for Psilocybes, because these grasses have a potential impact on the production of psilocybin and psilocin. Canary grasses have dimethytrypton content and they are also good habitats for psilocybin. It means that grasses have different impacts on growth and properties of Psilocybin mushrooms. Their environment has some morphological impacts which help to identify grassland psilocybin mushrooms easily. Some Psilocybin mushrooms from grasslands are from Sclerotia culture as they are hard in shape and have nut-like structures. Therefore some grassland psilocybin mushrooms are known to produce sclerotia which can survive environmental disasters, for example *P.mexicana*, *Conocybe cyanapus* etc. Most of the grasslands' species of Psilocybin mushrooms like moist and humid environments, and they are humus loving. These species grow in red clays or dark loams and are usually attached to tall grass so they are easy to pick. The grassland species of Psilocybin mushrooms are distributed and spread by different grazing animals like sheep, horses, cattle, yaks, water buffalo etc. Usually, these mushrooms remain undigested in an animal's body, and excreted through their feces, thus leaving these mushrooms to grow in dung places.

Dung Deposits

The species growing at grasslands and dung deposits share geographical niche. Most of the species which grow in grasslands also grow in dung deposits. As mentioned earlier, the animals who graze on grass excrete undigested spores of mushrooms in their feces and are responsible for the growth of Psilocybin mushrooms in dung deposits. Dung deposits are not permanent habitats: they are short lived habitats. So, the Psilocybin mushrooms grow there but cannot last for long time. *Psilocybe cubensis* is a dung dwelling species and its co-occurrence with a grassland species *Psilocybe mexicana*, is an example of relation of grasslands and dung deposit habitats. Other well-known species habiting dung deposits include *Psilocybe coprophila*, *Panaeolus cyanescens* and *Panaeolus subbalteatus*, etc. The dung of Cascade Mountains of Pacific Northwest are large habitats of Psilocybin mushrooms and provide a easy to find habitat for these mushrooms.

Burned Lands

Although burned lands are not very good for the growth of Psilocybin mushrooms, sometimes good fruiting can be found there. In some regions like Central Oregon, there is a practice to burn fields and grass-seed lands, then their re-growth. Such lands have the evidences of growth of fruiting of Psilocybin mushrooms, like *Psilocybe strictipes*. The natural rebirth of these lands supports the growth and development of Psilocybin mushrooms. The soil erosion, cracking, etc. can occur in burned lands. But, the growth of Psilocybin mushrooms can also be supported by flooding etc.

Riparian Zones and Disturbed Habitats

Riparian zone is the area which is a link or connection between land area and stream or river. They are mostly created by flooding in rivers. When flow in rivers is at its peak, it passes over the plants and tress, thus causing some damage but after that overflow, a riparian zone forms. These regions have high sand content but are a great source of biomass. Many species of Psilocybin mushrooms grow in these riparian zones because they find suitable environment there. For example, *Psilocybin quebecens* is a popular species which grows in riparian zones. The mushroom species which grow there also grow in disturbed regions. Disturbed habitats are those which are formed by natural disasters like floods, earthquakes, etc. Hot springs and Geysers are also a combination of riparian zones and disturbed habitats caused by different changes in nature. For examples, *Psilocybe Cyanofiberillosa* is a psilocybin species living in such habitats. Lets discuss disturbed habitats in detail.

Gardens

Gardens come under the category of disturbed habitats because of the continuous tilling process. They are excellent habitats for Psilocybin mushrooms. Gardens help these mushrooms by daily watering, soil amendments, and growth of plants. Examples of species growing in gardens include, *Psilocybe baeocystis, Psilocybe caerulescens, Psilocybe Stuntzii* etc.

Woodlands

Woodlands are mostly populated with coniferous trees and other subtropical plants. It is a very suitable habitat for the growth of Psilocybin mushrooms. The humid and moist environment and soil promote the growth of these mushrooms and many other species.

Moss Lands

Moss lands are another habitat of Psilocybin mushrooms as they are covered with Sphagnum. Sphagnum is a popular peat moss which can be a substrate for Psilocybin mushrooms. Moss lands are not very good habitats for these mushrooms.

Chapter Three

The Biology of Psilocybin Mushrooms

The outer look of Psilocybin mushrooms is the straight stem, open flat cap, wide and dark, brown in color. Inexperienced persons cannot properly identify the Psilocybin mushrooms, and they may mix them with other mushrooms. Many mushrooms are toxic and may lead to death; therefore, a layperson must not search psilocybin mushrooms without the help of an expert. Experts can easily identify these mushrooms based on their experience. The Psilocybin mushrooms are mostly identified by their gills, which are white in color. These gills have linings in which spores are present. These spores are responsible for the spread of these mushrooms through different dispersal procedures. Not all gilled mushrooms are Psilocybin mushrooms. Psilocybin mushrooms have brown or black spores inside the gills, and sometimes these mushrooms have bluish bruises.

It has been discussed earlier that Psilocybin mushrooms are saprophyte, like all other fungi. They have the same internal structure like other fungi with a cap attached to a stem having gills inside. The stem is attached to the volva, and volva stands on hyphae, the hair-like structures protruding out, and attaches this mushroom to the ground - these are called mycelium. Almost all fungi have this structure with little exceptions, mentioned before.

The Life Cycle of Magic Mushrooms

The hyphae of mycelia of magic mushrooms pass through plasmogamy which is a process where the cytoplasm from two different parent cells fuses, but leaves the nuclei infused. Means, cytoplasm combines, but nuclei remain haploid. Psilocybin mushrooms have haploid gametes, i.e. (n) instead of (2n). They form a diploid mycelium from a haploid parent. When the environment is suitable, this mycelium grows into mushrooms. When the karyogamy occurs in the cells, they result in the formation of gills. Karyogamy is a process in which the two unfused haploid nuclei fuse to form diploid nuclei. To understand the life cycle of Psilocybin mushrooms, keep these points in mind:

- Spores release and disperse through wind
- Find a suitable environment and germinate
- Mating
- Plasmogamy occurs
- Dikaryotic mycelium
- Gills form containing basidia

- Haploid nuclei
- Karyogamy occurs
- Diploid nuclei form
- Meiosis occurs (formation of basidiospores)
- Release of spores
- Cycle continues

With slight differences, all mushrooms go through this life cycle. The internal structure and chemistry of all mushrooms are the same with some differences. For example, Psilocybin mushrooms contain Psilocybe, and it is present in these mushrooms due to horizontal gene transfer. The biological and chemical properties of these mushrooms will be discussed further.

Identification of Psilocybin Mushrooms

Many people mix these with other mushrooms. To identify Psilocybin mushrooms, high expertise and experience are required as wrong identifications can lead to health dangers. Some people may confuse magic mushrooms with others, and when they eat them, they may face mental health issues because they don't know the proper way to use these hallucinogenic mushrooms. Some mushrooms are deadly poisonous and the collectors may collect wrong mushrooms instead of Psilocybin mushrooms. So, a mushroom collector must have keen observation and knowledge about psilocybin mushrooms.

Hofmann and his team were the first who identified the Psilocybin mushrooms successfully using the laboratory methods. The real identifications were made in 1958; before that, people used to identify these mushrooms only through their outer look and assumptions. Hofmann and his team members changed the game. They used *Psilocybe mexicana* to grow its cultures in the laboratory and then grow its fruiting bodies, sclerotia, and mycelium using these cultures. They also observed that the same psychoactive activity was present in both dried and fresh samples of *Psilocybe mexicana*. That was the time when the compounds behind the psychoactive activity and hallucinogenic properties of these mushrooms were discovered. Improper identification of mushrooms may lead to death, so one must understand the shape and structure of psilocybe mushrooms to avoid misidentifications. As these mushrooms have hallucinogenic properties, they can alter the human mind, thus producing psychological effects. These effects can be for a short time period or longer, depending upon the quantity of mushroom taken and the power of that specific type.

To identify magic mushrooms, one must consider these points:

- Brown colored mushrooms (Brown fruiting bodies)
- Having white gills underside
- Most of the psilocybe mushrooms have a ring around their neck, so consider that too
- Spore color (mostly purple-colored spores are of psilocybe mushrooms)

This is not the ultimate guide for mushroom identification, so if you find such mushrooms and you need to identify if they are magic mushrooms or not, check them in your laboratory to confirm. You can also use spore printing as an easy method to discover those mushrooms. Use a piece of paper, and put spores on them, then check the color that they leave on tissue paper. It can help to identify the psilocybin mushrooms. Thus, two important methods of mushroom identification are:

- Spore printing
- Bluing reaction

Spore Printing

The spore color is the most important thing in mushroom identification. If you are a mushroom collector, or you are in search of Psilocybe mushrooms for some purpose, consider their spores. Take the mushrooms and put them in polythene bags, but remember, don't tighten the bags because they have to breathe. If mushrooms become dry, the spores will change their color, which will not help in identification. So, use fresh and live mushrooms.

Open the polythene bag, take a mushroom and cut it. There is a proper way to cut those mushrooms. Choose flattened mushrooms to cut and separate their caps from the stem using a knife. Put the gills of the mushrooms (underside of the cap) on the paper or tissue paper and cover it with some glass to increase the level of hydration and decrease the chance of dehydration. Keeping the air currents away will give good spore print. If the gills are humid or wet, they will give a good spore print. If they leave purple or dark purple color on tissue paper, they are surely psilocybin mushrooms. This process is called spore printing because the spore leaves its print on the tissue paper for identification purpose. According to the symmetry of the gills, the spores will print on the paper, showing the mass and color of spores. After identification, you can label the print, and you can use these spore prints for mushroom cultivation in the future.

Bluing Reaction

The second important strategy to identify psilocybin mushrooms is their bluing reaction. Many psilocybin mushrooms have common properties that become the source of their identification. Most of the psilocybin mushrooms become bluish or bluish-green when they are bruised. Mushrooms do not only get bruises when they are crushed: bluish bruises can show when they are simply handled or picked. This bluing reaction shows that there is something different in these mushrooms. Scientists and researchers are in search of finding the reason behind this bluing reaction. One reason behind this bluing is the degradation of psilocybe and psilocin compounds. Some other unknown compounds also participate in this reaction. This is the most important identity of Psilocybin mushrooms as this property is not shown by other types. But, on the other hand, some psilocybe species don't show this blue color and no bluish bruises.

After some time, researchers also found that there are some poisonous mushrooms with bluish bruises, but they have no psilocybe. For example, the *Hygrphorus conicus* is a poisonous mushroom species that shows a bluish reaction when picking it up. So, now the bluish reaction is a primary parameter for Psilocybin mushroom identification, but not the least parameter. A trained person is needed to identify the correct species.Experts can also check the psilocybin mushrooms in labs to confirm the presence of specific psilocin and psilocybe compounds.

The Narrowing Down Technique for Identification

If someone wants to get expertise in the identification and collection of psilocybin mushrooms, he must follow the narrowing down technique. First, collect the mushrooms which you doubt are psilocybin mushrooms. Collect conical-shaped cap mushrooms, with a long stem, dark brown pointed caps, and brown gills. Place these mushrooms in plastic bags. Bring them to the laboratory and let them dry until the gills start to get yellow and white stems start turning brown. Place the gills on a paper and rub it with the paper to check the color of the spores. The purple spores appear, leaving a spore print. Check the mycelium; it will have blue bruises. Take a closer look at the edges of the gills and you will notice a white-colored band around it. If the mushrooms have all these characteristics at once, then they are from the psilocybe genus.

For further confirmation, you can also use a laboratory test to check the presence of psilocybe and psilocin. This technique will surely help to identify the psilocybin mushrooms. However, practice makes a person an expert in something, so don't trust your first observation for identification. Both techniques, like bluing reactions and spore printing, requires deep observations. If a person has good observation, he can become a Psilocybe identifier in a short time.

Why They Are Magic Mushrooms

Why are these mushrooms called magic mushrooms? Their magical properties are the reasons which lead to their use in rituals and religious ceremonies. The priests and mystics believed that these mushrooms have spiritual effects on the human brain. Therefore, many people started to worship these mushrooms because they thought that it has powers gifted by God or some spiritual powers. All these things are mentioned before in history etc. When science progressed, scientists and researchers started searching on Psilocybin mushrooms to find the mystery behind their magical properties. The research showed that these mushrooms were not magical and were not hallucinogenic long before. They got magical properties and hallucinogenic compounds by evolution.

There were lots of these mysterious mushrooms and insects used to come and eat them. Psilocybin mushrooms, like other organisms, developed such properties to protect themselves from insects. These mushrooms got Psilocybin and Psilocin compounds from the environment. The adaptation of these compounds was then in the genes of those mushrooms. Mushrooms got the genes of Psilocybin and Psilocin by horizontal gene transfer, thus gaining the hallucinogenic properties, and therefore called magic mushrooms. The psilocybin mushrooms developed the power when insects came near or tried to eat them; mushrooms altered the insect's mind, and the insects decide it is in their best interest to leave the mushroom as it is, refusing to consume it. This is the property that many organisms adapt to protect themselves from predators. We can say that mushrooms became Psilocybin mushrooms for their survival.

Scientists kept on researching on these mushrooms and their compounds. They are called magic mushrooms because they alter the minds and put some magical effects on the brain. Scientists and mycologists are still searching on different species of magic mushrooms. These can play an important role in treating different mental disorders, as depicted by the research. As these mushrooms alter the working ability of the brain in insects, it can also put some magical effects on the human brain. The compound Psilocybin has been studied for its use in the treatment of mental disorders. So, Psilocybin mushrooms can be used as a treatment of chronic depression and anxiety. It has been studied that the compounds psilocybin and psilocin present in these mushrooms have the ability to pull out the negative thoughts from the mind, hence the reason they can be helpful in the treatment of many mental disorders.

What Is Magical in These Mushrooms

These mushrooms have been successfully used in some mental therapies due to its components. The psychedelic compounds in these mushrooms were tested on patients with mental disorders. The results of the data obtained showed that those patients expressed positive changes in their attitude. The research reported that those patients treated with Psilocybin mushrooms were found to be closer to nature after treatment. Even they changed their religious and political points of view. Psychedelic research group (PRG), performed research by hiring volunteers having depression (which showed resistance to treatment). They were given oral Psilocybin (extracted from Psilocybin mushrooms), and they were also provided with little counseling during the research. Then a paper was published in this research in the Journal of Psychopharmacology. The research showed that the patients were connected more to nature, and their political views were also changed. It clearly depicts the strong connection of Psilocybin mushrooms modifying brain activities and thoughts.

This natural substance which changed the human mind in lesser time is considered strange. Researchers and scientists are still working on these mushrooms to find the exact phenomena on which these Psilocybin mushrooms work to create hallucinations. These mushrooms have a good taste, but when animals or insects eat them, it acts with the neurotransmitters of the animals — thus altering their nervous system. This is the phenomenon that makes these mushrooms magical, and that's the reason that mystics used them in their rituals and performances.

The Short Biological Properties of Psilocybin Mushrooms

Biological properties of Psilocybin mushrooms, extracted by the research are:

- Psilocybin mushrooms became hallucinogenic and magic mushrooms by horizontal gene transfer thousands of years ago.
- They have slightly different looks as compared to other mushrooms, but their identification is only possible with high expertise.
- The life cycle of Psilocybin mushrooms is the same as other mushrooms, and they are saprotrophs.
- Other animals and plants are affected positively or negatively by the presence of these psilocybin mushrooms.
- They used to grow in grasslands, gardens, animal dung, and in some disturbed habitats.
- The hallucinogenic properties of these mushrooms have made them unique.
- They come under the category of edible mushrooms, but not as a food, but as a medicine
- In many countries, the cultivation of Psilocybin mushrooms is not legal due to its drug-like properties.

- Psilocybe is grouped as Drug 1 category, because it puts hallucinogenic effects on the human mind.
- A person feels nausea after eating hallucinogenic mushrooms.
- These mushrooms like to grow in dark and humid places, therefore, old mushrooms are found in caves.

Their spores allow them to make themselves easily dormant when the conditions are inappropriate.

CHAPTER FOUR

Properties of Psilocybin mushrooms

Magic mushrooms first evolved by the effects of the environment. The changes in climate and environment bring changes in the behavior of Magic mushrooms. The main factors affecting these mushrooms include humidity, soil conditions, resource competition, disease impact, and surrounding predators. Predators, like insects, have been a major factor in bringing changes in Psilocybin mushrooms. The environment also affects how the magic mushrooms grow and the number of chemicals they have. As mentioned earlier, the major compounds present in magic mushrooms are Psilocybin and Psilocin. It is not necessary that all the compounds are present in equal amounts in all parts of the mushrooms. Different parts contain different amounts of these chemicals. There is a difference between the chemical properties of wild and cultivated mushrooms. The wild mushrooms show high variabilities in the chemical compounds they contain. The reason is, wild mushrooms have vast habitats, and they can grow in many different places, even in caves. While cultivated, mushrooms have only a few major chemical compounds because they are grown under controlled conditions at specific places.

Chemical and Physical Properties of Psilocybin

Psilocybin is water containing a crystalline compound. It is sometimes recrystallized from water. The soft, white colored crystalline needles, containing water, is an actual form of Psilocybin. This compound melts at 220 to 280 degrees. Psilocybin is 20 parts water, but if we dissolve it in methanol, it is soluble in 120 parts of methanol. In boiling methanol, thick prisms of Psilocybin are produced, which then form crystal methanol, the melting temperature of this crystal is 185 go 195 degrees. If someone wants to put the compound in some insoluble solution, then use chloroform and benzene. The solubility of Psilocybin is very poor in ethanol, according to experiments.

The Psilocin is actually the degradation product of Psilocybin. When boiled in methanol, the degradation product, Psilocin, forms which is almost insoluble in water, but it is soluble in some organic solvents. These compounds, when isolated by different methods, can be visualized by coupling them with some reagents. The Keller reagent (the iron chloride dissolved and concentrated in acetic acid and sulfuric acid), and the Van-Urk reagent (p-dimethyl benzaldehyde) are two common reagents which can be used to visualize psilocybin and psilocin by coupling. Psilocybin produces violet color while psilocin produces blue color, when they are coupled with reagents. These reactions were discovered by Hofmann and his team in 1958 and 1959. The chemical reactions show that these compounds are also present in nature, other than in Psilocybin mushrooms. We can relate the blue psilocin with the bluing of psilocybin mushrooms when they are picked up. These compounds have high value according to pharmaceutical point of view. The bonds formed by Psilocybin with hydrogen, in other compounds like LSD, can be studied through X-ray crystallography. The melting and boiling points of Psilocybin also depends on the solvent in which they have been dissolved.

The melting and boiling points of Psilocybin come under the category of physical properties, while its reactions and solubility come under the category of chemical properties. For their different purposes, scientists and researchers tried to synthesize the Psilocybin and Psilocin in the laboratory. These compounds can be synthesized in the laboratory as well as can be extracted from Psilocybin mushrooms. To synthesize these compounds, high expertise and understanding of synthesis methods are necessary.

Physical and Mental Effects of Psilocybin and Psilocin

The physical and mental effects of psilocybin mushrooms are interesting to discuss. The compounds present in these magic mushrooms, mainly Psilocybin and Psilocin, have amazing effects, depending upon the ingestion. If someone takes a number of magic mushrooms and keeps them in the mouth for 10 to 15 minutes, they start feeling something different. They start having some psycho-activity when swallowed. The effects of taking this amount of Psilocybin mushrooms are yawning, feeling some restlessness, and some bitter taste in the mouth.

Taking any amount of psilocybin mushrooms can cause some physical reactions. Some prompted physical reactions of Psilocybin mushrooms are the dilation of pupils, dry mouth sensation, rise in blood pressure, high rate of heartbeat, high temperature, etc. These effects are due to the inhibition of important neurotransmitters, serotonin. There is a similarity between LSD and Psilocin as they both acts on similar mechanisms.

The physical effects produced by ingesting Psilocybin mushrooms mostly don't happen immediately. It depends on the amount taken in the mouth. The high dose of Psilocybin mushrooms taken in the mouth produces the effects after 8 to 10 minutes and sensations start after 15 to 30 minutes when the chemicals absorb in the stomach wall. They take almost an hour to reach the brain and cross its barriers. When they cross the brain barriers, they start their psychoactivity. The first signs include: yawning, malaise, restlessness. Nausea is also caused by these mushrooms in some people. Mainly, the species *Ps.Caerulescens* and *Ps.Aztecoru* cause severe nausea. Some people feel weakness in the legs, discomfort in the stomach, and chilling effects. These conditions persist for a short time period. After that, most of the people feel physically light and easy. But, in some people, these conditions last for a long time. The studies of the ingestion of Psilocin and Psilocybin in animals showed high doses of these compounds orally. The oral ingestion results in the distribution of these compounds in the whole body. Kidneys get higher concentrations than other body parts.

Chapter Five

Medicinal Properties of Psilocybin Mushrooms

Psilocybin is a chemical extracted from 100 species of mushrooms that belonged to kingdom fungi. These mushrooms are cultivated as well as grown naturally. Psilocybin is a classic hallucinogen produced in 1958 that was used in spiritual mystic sessions to acquire special connection with spirit. It has medicinal effects used to treat alcoholism, anxiety, obsession and understanding schizophrenia. It is actually a potential psychiatric wonder drug in that period in the '60s. It was used before 1970 as a ritual to reduce depression. The Controlled Substance Act stopped the usage of psilocybin in 1970 in clinical studies using hallucinogens and psychedelics. Research on theses agent mostly was completed in the 1950s and 1960s, but it was not taken seriously because of the small nature of studies lack of professionalism and incompetence with current research standards. If studies on this had not been stopped, we might get a lot of wonders in the medical field. We can recreate new ways to treat mental illnesses and other life-threatening diseases. It can lead to the discovery of neurotransmitters and can calm down nervousness, sleeplessness and anxiety. It's probably a dedicated chemical to use as an antidepressant, which can help to solve the mental crisis. The National Institute on Drug Abuse, Food and Drug Administration advisory committee allowed resumption of research on psychedelics agents in 1992. It changed the perception of drug.

Biological Impacts

Psilocybin mushrooms, as shown by their name, are known for the presence of alkaloid compound psilocybin. Psilocybin has very strong **hallucinogenic effects**, so these mushrooms, if ingested, affect your body's metabolism to a great level. As soon as psilocybin enters the body, it is converted after metabolism into another compound known as psilocin, which starts to bring changes in the nervous system that induces hallucination. (Hallucination is a delusional state of mind in which a person sees or hears vivid things without any external stimulus). Psilocin is the primary psychoactive substance present in psychedelic mushrooms. It (psilocybin) produces the same effects as produced by other drugs like **mescaline** and **LSD** (lysergic acid Diethylamide). This is not life-threatening usually but, due to its strong hallucinogenic characteristics, it may produce very unpleasant effects like other hallucinogenic substances. However, there are some risks that are specifically related to mushrooms. The biological impacts of magic mushrooms can be seen physically just after the intake, but these mushrooms also produce alteration at cell or tissue level, and these internal effects may not be felt at once but appear later with the passage of time.

Physical Effects

The physical effects of psychedelic mushrooms are,

- Dizziness
- Increased heartbeat
- High temperature of the body
- Headache
- Weakness in muscles and body
- Increased blood pressure
- Sleepiness

Mental Effects:

- Euphoria
- Auditory or visual Hallucination
- Neuropsychiatric instability
- Distortion in sensory perceptions
- Paranoia
- Psychosis
- Panic reactions

Overall Biological Effects:

Some biological impacts are some symptoms that appear immediately or slowly after ingesting psychedelic mushrooms:

- Psilocybin attacks the parts of the brain that are associated with emotions like fear, rage, anxiety, etc. and develop stress and **anxiety.**

- Panic attacks are also observed after intake of psilocybin.
- The person feels detached and is surrounded by negative feelings and **impaired judgment** etc.
- If some wrong type of mushroom is ingested mistakenly, then it may cause severe poisoning and even death.
- **HPPD** (Hallucinogen Persistent Perpetual Disorder) can also occur in which there are flashbacks or reoccurrence of the effects caused by psilocybin long after the intake. Such a condition can be very unpleasant and distressing.

The National Institute on Drug Abuse illustrates all properties and biological impacts of every kind of hallucinogenic chemicals and drugs, including the psilocybin mushrooms that are also being abused for hallucinogenic and euphoric effects. According to this institute, the psilocybin mushrooms induce hallucination by passing through the **neural pathways** and acts on the frontal lobe of the forebrain cortex (that is primarily responsible for the actions and responses associated with vision, hearing, and smelling).

The compound psilocin performs its function by acting on the sensory parts of the brain with the help of the neurotransmitter serotonin. The action of psilocybin on the nervous system is not temporary; rather, it produces many long-lasting changes in physical as well as mental health. These changes are either positive or negative, depending upon the method of use, potency, and activity of psilocybin, the quantity of psilocybin, and the frequency of its intake.

Hallucinogenic Effects of Psychedelic Mushrooms:

Hallucination is a delusional state of mind that may be auditory or visual in which the sufferer hears or sees the things which do not exist in actual. Hallucination is triggered by several nervous disorders like Parkinson's disease, schizophrenia, brain tumor, and also by severe diseases of kidney and liver. It also occurs with the use of addictive drugs, such as marijuana, heroin, alcohol, cocaine, and nicotine. Psilocybin performs the main action on the nervous system and causes drowsiness and hallucination just as done by LSD and mescaline, but it has less hallucinogenic properties than these strong hallucinogenic drugs.

When the psilocybin from the magic mushrooms gets into the blood of a person, a metabolism takes place, which converts it into another substance, psilocin. Psilocin goes directly to the nervous system through blood circulation and starts its job as a psychoactive agent. It makes the user suffer from all those conditions that are caused by addictive drugs for example:

- bad trips
- Dizziness
- Unconsciousness
- delayed headache
- nausea
- drowsiness
- diarrhea
- muscle weakness and pain

- anxiety
- fear or rage
- psychosis

All these conditions arise from the excessive intake of psilocybin. The exact situation of the psilocybin user and the immediately appeared symptoms depend on the dose of psilocybin consumed by that person. Sometimes, a small dose does not have many negative effects, and also, they do not remain for a longer period of time to affect the body's health to a dangerous level.

But, if psilocybin is consumed frequently or taken in a large amount at one time, it may prove lethal to health. More quantity of psilocybin causes intense neurological and psychiatric disorders such as:

- The patient may go in a state of coma for a long time.
- Intense level of stress and anxiety
- Paranoia
- Hallucination and poor brain function
- Seizures
- Depression
- Vomiting and other digestive disorders

Psychedelics produce a number of long-term effects on the entire body of the consumer. Magic mushrooms were banned and were illegal to cultivate due to these hallucinogenic and also addictive properties as people can abuse them. A person may become dependent on it and subjected to severe panic conditions, least tolerance, and negative emotional changes.

Medical Uses

Besides having negative impacts on human health, psilocybin mushrooms also play a positive role in the medical treatments because of their magical properties. The medical treatment of certain ailments involving the use of psilocybin extracted from the psychedelic or psilocybin mushrooms is now getting popular and is known as "Psilocybin therapy." Psilocybin, the main effective ingredient of magic mushrooms, plays a great therapeutic role in the cure of many diseases; thus, it's a worthy topic of discussion and practice in the field of medicine. As the research and studies on magic mushrooms began, its first use was made to treat anxiety and associated disorders. Psilocybin mushrooms are significantly helpful in treating the stressful situations that arise in a patient when he/she is suffering from a chronic disease like:

- Cancer in any part of the body
- Trauma or injury due to some accident
- Some severe viral or bacterial infection
- Physiological or physical disorder.
- A serious distressing condition can also occur due to acute pain after a surgery

- Any irritation or panic attack during or after a treatment that involves frequent injections, stitches, high potency medicines, etc.

In such cases, psilocybin proves really helpful to relieve stress and pain. The dosage of psilocybin given to a patient is also very important for getting the required results in the treatment. It is usually given in the form of an oral capsule that contains 200mcg (given per kg weight of the patient) active psilocybin or 250mg niacin. Niacin is used as a control by scientists as it produces a warm flushing effect in the body; this is a common adverse effect of psilocybin, without producing an alteration in the brain or psychological state. It has been subjected to a number of experiments and was given to the patients at regular intervals (mostly 2 weeks) to prove its remedial abilities and found it to be remarkably effective in reducing anxiety. This therapeutic role of psilocybin is confirmed by different **research institutes,** for example, National Institute on Drug Abuse, Emma Sofia in Norway, Multidisciplinary Association for Psychedelic Studies (MAPS) in California, The Kings College London and the Heffter Research Institute in America.

Psilocybin works with the blood flow in the nervous tissues of the body. An experiment was done to see the role of psilocybin in reducing stress and anxiety by comparing the **MRI reports on cerebral blood flow** before and after a regular dose of psilocybin. It was observed that psilocybin decreases the blood circulation in those sensory areas of the brain, which are responsible for producing stress and anxiety. The most effective results of psilocybin therapy are seen after one week.

Psilocybin therapy is performed in a single session or a series of sessions at regular intervals, depending upon the severity of the medical issue and particular biological conditions of the patient. But, in most cases, a single session of psilocybin therapy is effective enough to give long-term cure to the patient.

Safety

Psilocybin has similar effects to lysergic acid diethylamide (LSD) and mescaline, but psilocybin is not addictive in nature: patients can easily stop its usage. Studies on a group of people showed that patients were safe from drug abuse, persisting perception disorders, prolonged psychosis, or other long-term deficits in functioning. The adverse effect of psilocybin was reported very few and mostly related to high doses. Experimental groups were observed for 8 months after the administration of the drug (psilocybin) and no harsh side effects were observed. People from the psilocybin administered group were calm and peaceful during all the experimental conduct. There have been some concerns regarding the use of psychedelics agents that they may cause mental retardation and cause an attempt of suicide among people who used them. A study conducted on people to observe those effects does not support the above claim. No significant relation was found between the use of psilocybin and mental retardation or suicidal thoughts, attempts or plans. It is the safest drug out there. For example, in 2018, just 0.3 % of people who reported taking them needed medical emergency treatment

Release Depressed Mood Swings

Studies showed that use of psilocybin could reduce anxiety suicidal thoughts and depression. The study was conducted using data from the national survey on drug use and health. They divided the participants into four groups who used psilocybin; only one used psilocybin with psychedelics, while the other was non- psilocybin psychedelics only and one group which used none of them in their entire life. Results shocked all the odds and myths against psilocybin. They were quite progressive and satisfactory about the use of psilocybin. The group that used psilocybin showed less distress and disappointment as compared to the other groups. They appear far better than any others. They showed improvement of moods and stress relief. MRI imaging reveals reduced blood flow in amygdala, a small almond-shaped region of the brain known to balance emotional responses, sleep-wake cycle and thinking process in humans. People started to take minor doses of it; they think it can improve their insight, perception of understanding of others, enhance a person's capacity for self-transcendence, which is needed to overcome self-deception. *It can make you more attentive and able to focus problems.* From the results, scientists conclude that psilocybin may play a role in decreasing anxiety, mood swing and bring peace in people's behaviors. Some treatments for mental health and care suggest the use of psilocybin. In certain cases, extreme use of this chemical may cause vomiting, nausea and hallucination, but this happens in rare cases with psilocybin abuse.

Tobacco Cessation

Psilocybin has a high affinity for several serotonin receptors like 5-HT_{1A}, 5-HT_{2A}, and 5-HT_{2C}, which located in numerous areas of the brain, including the cerebral cortex and thalamus. Studies have shown positive evidence that use of 5-HT_{2A}o receptor agonist is helpful in the treatment of addiction. Research was conducted on 15 participants enrolled in a smoking cessation course that includes administration of psilocybin in their therapy period at weeks 5,7 and 13. All other medications were stopped for those patients. They smoked 10 cigarettes per day. Cognitive-behavioral therapy had been done for a 4-week period while using cigarettes. The study revealed positive effects of smoking prohibition on patients. The group of people using psilocybin stopped smoking at the 6th week of study. This shows that psilocybin has a magical effect of releasing any kind of addiction. It is the best sensational drug to stop smoking among patients taking vernicles.

Psychological Effects

It has a calming effect; it also has a pathogen effect similar to MDMA, a helpful and beneficial compound in reducing stress and making good perception about problems and people. Psilocybin converted into psilocin which binds to serotonin receptor called 2A, and experts think that's what triggers what they call neuronal avalanching. It can cause different changes in the brain and you have increased activity in the cortex which leads to change in perception. There is also decreased network activity in the default mode network, which leads to loss of ego and that's why people report a profound sense of unity transcending beyond themselves. Psilocybin increases connectivity among different parts of the brain. Due to this receptor activation, different areas of the brain synchronize with each other like an orchestra. Once psilocybin enters, it coordinates and communicates all parts of the brain, also those parts which are normally compartmentalized and doing their own work. Scientists believe that it's a combination of these effects that makes psilocybin useful for combating addiction and depression. When new areas or region in the brain start talking to each other, it can be profound. Healing can be helpful in the thought process and give a critical analysis of problems. Despite all benefits, psilocybin is still listed as a Schedule I drug, a category reserved for compounds that have currently accepted no medical use.

Obsessive-Compulsive Disorder

A study has reviewed the potential benefits of psilocybin on obsessive-compulsive disorder. It is a mental health disorder that affected people of all ages and every field; it occurs when a person gets caught in a cycle of obsessions and compulsions. This can affect the prefrontal cortex. Patient with OCD found that prefrontal cortex gets relieved after the few weeks of usage. It is safer than alcohol, tobacco, cannabis, and is non-addictive. Studies showed a positive effect of psilocybin on this disorder. People who suffered from this disorder get rid of disease while using psilocybin. A study conducted on 9 patients with OCD, to check the impact of psilocybin mushrooms. They hypothesized that the oral administration of psilocybin would reduce the effects of OCD. Before the experiment, those patients tried the treatment for the disease.

Patients were given four different doses at least for one week. From among all the population, 88.9% of people showed a decreased effect of the symptoms.

Alcohol Dependence

Psilocybin is a magical drug used in mental health care. Its high affinity for serotonin 2A receptors in the brain will reduce addiction of many kinds. It can be helpful for people struggling with alcohol addiction. This is also proved through cognitive study. These are natural hallucinogens that is structurally similar to serotonin and DMT. Psilocin is an active biological form of psilocybin. There is a long history of the usage of psilocybin as a therapy for alcohol addiction. Now, in the current study conducted on that, people reported a very positive effect of psilocybin on alcoholism treatment. They reported mystical transforming experiences.

Anxiety Disorder

In the society we are living today, we need consistent fixers for our deep-rooted problems. From pharmaceutical medication to anti-depressant to antipsychotic meds, we are looking for a quick solution to bandage our sufferings. Results of these medicines may be quick, but they can affect people in many other ways. They can lead to long-term conflictions with existing problems. What if I can tell you a better alternative? These are the magic mushrooms from the backyard of your garden. These mushrooms have been used for thousands of years for almost every field.

Anxiety is an emotional feeling characterized by stress and tension. Different factors can increase anxiety in a person like the failure of losing something feeling to get success in life. Some patients with chronic and life-threatening diseases also suffer from anxiety like cancer. Studies revealed psilocybin reduced anxiety among cancer patients when administered with successive control dosage of psilocybin. Psilocybin can be used as a medication for the management of anxiety disorders. Magic mushrooms cure neurological disorders like PTSD, depression, OCD over active amygdale and migraines. Magic mushroom may reset the activity of the brain circuit known to play a role in depression. Use of magic mushroom can bring a very beautiful feeling. It has healing powers and has been used from many thousands of years to bring unity and enlightenment.

Cancer Therapy

Psilocybin has been used in rituals for centuries. Modern study revealed the medicinal effect of psilocybin in cancer treatment. Cancer is a life-threatening disease. Patients lose their all hopes for life; they go into a critical condition obsessed with dark thoughts. This negative approach of patients may adversely affect the immune system of the patient, which is already compromised. In some cases, the patient collapses due to the stress of the disease. Here, psilocybin again saves them from depression and anxiety. It can save a struggling patient's life. Its effect on the serotonin receptors reduces depression.

Reported by Johns Hopkins, upcoming studies will evaluate the use of psilocybin as a new therapy for opioid addiction, Alzheimer's disease, post-traumatic stress disorder (PTSD), post-treatment Lyme disease syndrome (formerly known as chronic Lyme disease), anorexia nervosa and alcohol use in people with major depression. A focus on precision medicine tailored to the individual patient is expected.

However, even if approved by the FDA, psilocybin would have to be reclassified by the DEA to schedule II substance for it to be available for patients.

How Psilocybin and Psilocybin Mushrooms Are Sold

Psilocybin mushrooms are used by many people despite the restrictions and specific laws. People who know the benefits of Psilocybin mushrooms always look for its doses. They can be used in medicines of anxiety and depression, which help to reduce the conditions. They are also used for complete treatment of many mental disorders. Some researchers grow these mushrooms on small levels to meet their research requirements while some people grow them at large scales for other purposes. Large scale production of Psilocybin mushrooms is allowed in some regions.

All the disputes of Psilocybin mushrooms are due to their hallucinogenic properties due to the presence of Psilocybin compound in Psilocybin mushrooms. So, the compounds Psilocybin and Psilocin are considered hallucinogenic and magic compounds due to their unique properties. According to a conference in Vienna held in 1987[1], Psilocybin mushrooms have been eliminated from Schedule A category of drugs. These mushrooms have drug-like properties as well as they leave hallucinogenic effects on the human mind and body. They are sold in dried form or in the form of powder in many regions. The powder form of these mushrooms seems like a drug.

Some professionals also extract Psilocybin compound from mushrooms and sell that compound in crystallized or powdered form. Such buying and selling are done in the black market because the selling and buying of this compound is totally banned because it is a real drug. The use of natural Psilocybin mushrooms, in low or high doses, cannot be restricted totally. It is because naturally grown psilocybin mushrooms cannot be stopped to grow. Nature doesn't allow people to become a drug addict, so there is some legal purpose behind the use of Psilocybin mushrooms. The different forms in which these psilocybin mushrooms are sold are:

- In the form of capsules
- In the form of micro pills
- In the form of powder
- In the form of solid dried mushrooms
- In the form of extracted psilocybin

These mushrooms are sold mostly in the black market in the above-mentioned forms. Fresh mushrooms can be picked up easily, but most of the people are not sure if they are picking up Psilocybin mushrooms (liberty caps etc.), or they are just picking the death caps.

Use of Psilocybin mushrooms should be safe; therefore, you must take care while using these mushrooms. If you are using Psilocybin mushrooms or their capsules, you must not be alone, so if you fall, someone is here to pick you up. The small amount and doses of Psilocybin mushrooms can help you to get rid of negative thoughts and reduce anxiety slowly.

Chapter Six

Some Popular Psilocybin Mushrooms of the World

Many regions of the world are rich in Psilocybin mushrooms. These mushrooms grow naturally, so no one can put a ban on their natural growth. The countries that have restricted the cultivation of Psilocybin mushrooms due to several reasons cannot restrict the natural growth of these mushrooms in marshlands, gardens and other suitable places. So, mycologists, mushrooms lovers and researchers in addition with mystics, always search for Psilocybin mushrooms, identify them and patent them. Every year, they discover a new species of Psilocybin mushrooms and patent it with the name of discover. Let's discuss some popular Psilocybin Mushrooms and their genera.

The Genus Panaeolus

The genus Panaeolus also comes under Psilocybin mushrooms. Some popular species of this genera are:

Panaeolus Africanus

The *Panaeolus Africanus* belongs to the genus Panaeolus and has a conic cap with convex and hemispheric shape. Its surface is smooth but may have cracks or scales above the cap, these cracks and scales show up when these mushrooms are exposed to the sun. They are viscid when wet, mostly when they are young. The color of these psilocybin mushrooms is grayish, creamy white mostly but sometimes they have reddish-brown color but become grayish with age. Like all Psilocybin mushrooms, these mushrooms also have gills which are attached inside the cap, widely spaced but with an irregular arrangement. They look grayish at first look and become black with age, especially when they produce spores. They stem of these psilocybin mushrooms is 30-50 mm by 4-6 mm thick, firm, protruded towards the apex and equal. The color of the stem is mostly white to pinkish tones; the color of the stem is lighter than its cap.

These mushrooms are mostly found in the region of central Africa and in some regions of Sudan. During the rainy seasons, the mushrooms are found even on elephant dung in these regions. As discussed earlier, these mushrooms also contain Psilocybin and Psilocin; therefore, these mushrooms are Psilocybin mushrooms.

Panaeolus Castaneifolius

Panaeolus castaneifolius which is also known as Murrill regionally is a native Psilocybin mushroom of North and South

American Countries. These mushrooms have a spherical cap which becomes convex with age, curved margins at an early age and then starts to straight with age. The color of this species of mushrooms is smoky gray when they are slightly moist and when they start to dry, the color becomes straw yellow or pale. They are reddish-brown at the edges and margins. They are also gilled mushrooms, and their gills are attached to the inside face and wrinkled to some extent. The color of gills become purplish, gray-black when spores start to mature. The stem of these mushrooms is long and this but narrow at the base. It is gray in color, hollow and tubular. They are identified by their dark, purplish, gray black gills which are gray-black due to their spore's maturity.

They are widely distributed as mentioned before. The South and North American region are rich in this species of Psilocybin mushrooms as discovered by the researchers. They specifically grow in dark places.

Panaeolus Papilionaceous

The cap of this *Panaeolus papilionaceus* is conic in shape which becomes campanulate with age, the margins of its cap have a tooth-like structure in white color. In young fruiting bodies of these mushrooms, the surface of their caps is smooth but horizontally wrinkled having light-colored flesh. The flesh is thick on the underside of the gills. The gills are also attached inside the cap, which is broad to some extent and grayish in color. The grayish black color or gills is mostly unbalanced ripening of spores. The stem of these mushrooms is about 60 to 140 mm long and 1.5 to 3.5 mm thick. It is equally distributed, tubular, fibrous and protruding towards the apex slightly. The mix of brownish and grayish color makes these mushrooms differentiable from other mushrooms.

The whitish tooth-like structures look like a veil of these mushrooms. These mushrooms mostly grow in dung and in the season of fall or spring. They are native to North American and temperate regions of the world. Researchers noticed that these mushrooms look common during the autumn season. These mushrooms also resemble with some other mushrooms like *Panaoelus retirugis* etc. due to the whitish, tooth-like structures. Researchers have found that these mushrooms are identified by the presence of wrinkled cap.

Panaelus Semiovatus

Panaelus semiovatus is a psilocybin mushroom species with conic cap expanding near the convex. The young fruiting bodies are pink in color but become whitish with age. The smooth and wrinkled surface of its cap is due to the environment in which they grow. Whitish gills are attached downside of the cap, but they become brownish in color and then blackish when spores start ripening. They are mostly found in South America and Temperate zones of Europe. According to some studies, this species of psilocybin mushrooms is considered to be non-active for years. It mostly grows on dung. So, researchers looking for this species of mushrooms, mostly look in dung to find these species. It is identified by its viscid cap and large size of its cap as compared to other species.

The other popular species of Psilocybin mushrooms categorized under the genus Panaeolus are:

- Panaeolus antillarum
- Panaeolus cambodginiensis
- Panaeolus castaneifolius
- Panaeolus cyanescens
- Panaeolus fimicola
- Panaeolus foenisecii
- Panaeolus papilionaceus
- Panaeolus subbalteatus
- Panaeolus tropicalis

The Genus Psilocybe

Psilocybe is the most popular genera of Psilocybin mushrooms. They are also mostly called "Psilocybes." Like other mushrooms they are also saprotrophic and get their food from other organisms, they grow mostly in moist places, and their habitat is decaying wood debris, dung, grasslands, mosses etc. All the psilocybes contain Psilocin and Psilocybin compounds which separate the psilocybin mushrooms from other families of Mushrooms. They are identified by their brownish gills with white edges. Good expertise is highly recommended to identify the Psilocybin mushrooms. Some popular species which comes under the category of Psilocybes are:

Psilocybe Aeruginosa

The cap these *Psilocybe Aeroginosa* is convex to campanulate. The cap starts to expand as convex with the passage of time. It shows the shape like low, broad umbo. It is mostly bluish-green at the start, but the color fades away with age. The surface of the cap is mostly viscid due to moisture and humidity. It has margins covered with veil-like flecks. The gills of these mushrooms are brown in color, mostly clay brown, sometimes show a shade of purple and white edges. The stem is thick but swelled at the base. The surface of the stem is covered by white-colored patches. These mushrooms are mostly found in the regions like the British Isles, northern Europe, and in western North America. The perfect habitat for these mushrooms is wood debris, gardens, parks, grassy area at the edges of woodlands, etc. In the region of Pacific Northwest, the Psilocybe aeruginosa grows under the conifers while in Southern California, these mushrooms are found growing under the oak trees. Historically, these mushrooms are reported as poisonous due to enough content of Psilocybin.

Psilocybe atrobrunnea

Psilocybe atrobrunnea is the species of genus Psilocybe which has conic caps, but bluntly conic caps. They have sharp nipple above the cap and are convex with age. The caps of these mushrooms are mostly reddish-brown or blackish reddish brown.

When they get dry, their color converts to pale reddish-brown. The cap has a smooth surface, but viscid when they are moist. These mushrooms also have whitish veil-like structures. The color of gills of these mushrooms is dark purplish-brown and has whitish edges. They have a thick stem with a swelled base and equally distributed. The spores of these mushrooms are violaceous, dark brown in color and become black with the passage of time.

These mushrooms are usually found near sphagnum bogs, growing under conifers, woodlands and their fruiting start in September and October mostly. This species of Psilocybin mushrooms is found in Michigan, upper New York regions of United States. They are also reported from British Columbia and central to northern Europe (including Great Britain, Czech Republic, Slovakia, Finland, France, Germany, Sweden and Poland). It shows that these mushrooms are widely distributed. This species also has resemblance with some other Psilocybe species.

Psilocybe Aucklandii

The cap of *Psilocybe aucklandii* is conical in shape when they are immature and starts becoming plane with maturity. The margins of this species are striate and split with the passage of time. This species has no veil-like structures. The color of its cap is dark brown, but when it dries, the color becomes pale yellow or straw brown. It is discussed earlier that most Psilocybin mushrooms show bruises and become bluish when injured, same is the case with Psilocybe aucklandii. They show bruises and start to show a blue color when they are injured. The gills of these mushrooms are attached under the cap and are grayish-yellow in color and become darker with maturity. The thick stem of these mushrooms is covered with silky white fibrils, have brownish flesh and bluish bruising. The spores are purple colored and are distributed and scattered by different means. Psilocybe aucklandii are distributed naturally, and they like to grow on the soil rich with wood debris.

The name Aucklandii belongs to the Auckland city New Zealand. They are only found around the New Zealand, near the pines and woodlands. During the transport of woods, trees etc., these species have high chances to transfer to other temperate zones of the world.

Psilocybe Mexicana

This species of mushrooms is one of the most popular species. It is the one which was used by old mystics and was a part of religious rituals in ancient times. The cap of this strange species of mushrooms is conical to campanulate and becomes convex at maturity. The margins of its cap have fine fibrils, and its color is brown to deep orange which fades away on drying and become yellowish. The gills are attached inside the cap which is pale gray to dark purplish-brown on maturity, and edges become whitish. The stem of *Psilocybe mexicana* is thick and long with straw yellow to brownish color and becomes darker with age or any injury. The region of Mexico is rich in *Psilocybe mexicana*, and it mostly grows in horse pastures, soils rich with manure and fields.

The first research on Psilocybin mushrooms was started by *Psilocybe Mexicana*. Mystics and healers were using this mushroom, but they didn't know about their name and exact properties. This psilocybin mushroom was also worshipped by old people as they believed that this is a powerful mushroom. They think that this mushroom has some powers from God.

Psilocybe Cubensis

Another most popular Psilocybin mushroom from Psilocybe genera is "*Psilocybe cubensis*". It is one of the most commonly available psychedelics. There are some other names of this Psilocybe including *Stropharia cubensis, Stropharia cyanescens and Stropharia caerulescens*. It is also known as "Mexican mushroom" and magic mushroom or shroom. As compared to other mushrooms, this mushroom is easy to grow. But thanks to nature, this magical mushroom grows naturally most of the times. It also contains Psilocin and Psilocybin which are responsible for many of its medicinal, and hallucinogenic properties. Like *Psilocybe mexicana*, it is also the psilocybin mushroom which played an important part in research on Psilocybin mushrooms. It is also known as golden tops because of its golden colored cap.

The cap of *Psilocybe cubensis* is conic and campanulate, which becomes convex at the edges. Gills are attached downside the cap and are grayish brown in color which become purplish-black with age. The stem is thick and whitish but bruising may appear even due to minor injuries. This species of mushrooms is found in the southeastern region of United States, Mexico, Cuba, Central America, northern South America, in the subtropical Far East (India, Thailand, Vietnam and Cambodia), also in the regions of Australia. It is mostly seen in two months of its fruiting like in May and June.

Chapter Seven

The Basic of Cultivation

The term magic mushrooms allude to the fruiting groups of certain fungi. They contain psychoactive mixes, for example, **psilocybin** and **psilocin**. There is a wide range of psychoactive blends in magic mushrooms. The blends present change from species to species and groups to clusters. Along these lines, making careful and exact doses of psilocybin from dried magic mushrooms is troublesome. The mixes present in magic mushrooms cooperate in an assortment of approaches to make various impacts for the client.

Parts of a Mushroom

Growths are one of a kind life forms with body structures and regenerative modes not at all like those of some other creature. Mushrooms, shape and certain fungi are on the whole growths. The key highlights of a contagious body are the mycelium (comprised of hyphae), the fruiting body and the spores.

Highlights - Many growths look like plants, yet fungi are heterotrophs, similar to creatures. A fungus must process food to live, while plants are autotrophs that make their food through photosynthesis.

Mycelium:

A contagious mycelium is a system of threadlike fibers called **hyphae**. The mycelium acquires supplements (as a rule from rotting regular issue) and delivers the fruiting body. Frequently the majority of the mycelium will be underground. The mycelium of one mammoth fungus growing in Oregon covers more than 2,200 sections of land of woods.

Fruiting Body

The fruiting body of a fungus is a regenerative structure. A mushroom is an ordinary contagious fruiting body, appended to a mycelium underground. A fruiting body produces spores.

Spores

Spores are engaged with widespread propagation. Discharged by the fruiting body, contagious spores are haploid, which means they convey just a single chromosome for every quality (like human gametes). Spores can develop when they strike sodden soil.

In contrast to creatures, growths don't process food inside. They discharge stomach related chemicals, so their food is

"processed" outside their bodies. A fungus at that point obtains its supplements by retention of the processed food through the mycelium.

The Most Effective Method to Identify Wild Psilocybin Mushrooms

Wild psilocybin mushrooms are found in numerous areas over the globe and happen in any event, ten unique assortments. The most well-known of the wild psilocybin-containing mushrooms, **Psilocybe Cubensis,** is found in the United States, Central and South America and the West Indies. Psilocybin mushrooms can frequently be perceived by their shading, shape and stem wounding, which creates a blue shading. Psilocybin mushrooms ought to be deliberately kept away because they convey significant wellbeing dangers and are illicit to use in the United States. The ingestion of these mushrooms can cause mind flights, sickness, retching, sleepiness or even renal disappointment. Continuously cautiously recognize any mushroom being referred to before utilization to guarantee that it isn't of this sort.

- Look at the shade of the mushroom. Youthful Psilocybe Cubensis mushrooms (ordinarily these will be littler ones) might be a profound, brilliant earth-colored shading, while increasingly develop ones are of a lighter coarse colored shading.
- Look for an inside stamping. The Psilocybe Cubensis has an unmistakable darker earth-colored spot in the focal point of the mushroom.

- Check out the stem, whether there is a blue tone. This shading, which might be brought about by an oxygen and psilocybin connection happens with the wounding of any kind. If the mushroom has been moved by a human, bug or even grass or different mushrooms, this response is probably going to happen.
- Look for a profound purple shaded gill spread. This mushroom cloak is a goody covering that stays on the mushroom gills until the mushroom top completely extends, so all things considered, it will break. A messed up shroud can frequently be watched revolving around the stems of psilocybin mushrooms.

The Lifespan of Theses Fungus:

Fungi have an extremely short life expectancy. However, it contrasts extraordinarily from species to species. A few kinds may live as short as a day, while others endure anyplace between a week and a month. The existence cycle of a fungus starts as a spore and goes on until germination.

Spore Development:

Growths start their life as spores that are released from completely developed fungi. The cells of the spore separate and evolve into hyphae after there are discharged. When hyphae shaped from spores discharged by two distinct growths meet, they may intertwine to make a solitary cell with two cores.

Mushroom:

When the two-core cells, likewise called **dikaryon**, have developed, they form into a fruiting body known as a mushroom. The cores of the cells in the mushroom will experience further divisions and, in the long-run structure, haploid cells - cells with a solitary arrangement of chromosomes - that at that point isolate through meiosis and make spores.

Death:

When a fungus has finished making spores through meiosis, it adequately kicks the bucket. The spores spread and the rest of the tail and hyphae don't get past another fruiting procedure. The remainders of the fungus at that point deteriorate in the dirt.

Life Cycle of Psilocybin Mushroom

At the point when you start with the cultivation of mushrooms at home, it is imperative to think about the lifecycle of the mushroom. To effectively develop your mushrooms at home, it is necessary to have an understanding of the lifecycle of the mushroom. The mushroom exists of a stem and a top. What these individuals don't understand is that there's a whole system of alleged mycelium underneath the mushroom. Mycelium is a tight system of cells under the ground. This underground mycelium is the plant of which the mushrooms are the natural products.

During its life course, the mycelium has just a single **objective:** The current of the species. The mycelium does as such by growing mushrooms. These mushrooms produce spores and drop these when they arrive at adulthood.

The Lifecycle

- A developed mushroom drops spores
- Spores fall on the ground – spores develop
- They meet perfect spores
- Mycelium starts – pinheads begin
- Primordia arrangement – development mushroom – develop, drops spores
- Circle finished

By and large for the cultivator this way too:

- Obtain spores
- Make substrate
- Inoculation
- Incubation
- Place them into fruiting conditions
- Pinhead arrangement
- Grow the mushroom
- Make a spore print

Reproduction of Mushrooms

Spores

Mushroom trackers manufacture through soggy lush regions looking for the valued tasty mushroom. They should be talented in the recognizable proof procedure since certain mushrooms are lethal. There are more than 3,000 types of mushrooms all through the world. It is a fungus and not at all like different plants, has no chlorophyll to assist it with assembling food. The top of the mushroom - the part we commonly eat - is the fruiting piece of the mushroom and is imperative to its proliferation procedure. The top will last just a couple of days. However, during that time, it will make a huge number of spores. Spores are single cells, each equipped for forming into a mushroom.

Cells Produce Spores:

The cells that produce spores on the fruiting body are either **asci** or **basidia**. With asci cells, spores are created inside and in the basidia, they are delivered remotely. Spores are discharged when either the tip of the asci severs or the spores from the basidia. After the spores are discharged, they are conveyed by the breeze and it is feasible for them to land a long way from the parent mushroom. After the spores are discharged, the top or fruiting piece of the mushroom bites the dust.

Asci and Basidia Cells

The asci cells are situated on the inside surface of the cup organisms. At the point when the asci tears open, the spores are discharged. Gilled mushrooms, boletes and puffballs all have basidia cells. In the gilled mushrooms, they are situated on the base of the top, from which the spores are dropped. In the boletes, they are situated in tubes enclosed in the tissue of the mushroom top, with pores that discharge the spores. In the puffball, they are situated in the body of the top and the spores discharge when the shell of the top tears open.

The Cycle Continues

For a spore to endure and develop into another mushroom, it must land in a situation that is fitting for mushroom cultivation. The dirt ought to be moist and clammy. Mushrooms flourish in regions that are verdant and lush. After arriving in such a situation, the spore will develop hair-like fibers that are called **hypha**. From the hyphae, the mushroom's mycelium will develop. It is the piece of the mushroom that develops underneath the dirt. From the mycelium, a tailor stem will develop and on the stem will develop the fruiting top. At the point when the hypha of one spore meets with the hypha from another spore, a mating or germination process starts that outcomes in the creation of more spores.

Chapter Eight

The Basic Growing Parameters

Psilocybin mushrooms are probably the least demanding item to develop on the planet — they need a couple of explicit parameters and a touch of persistence. The units that we sell accompany a perlite and vermiculite substrate, which has had mycelium added to it, which is the place the shrooms originate. It's quite simple to initiate and start growing your shrooms; continue perusing for a full guide on the most proficient method to do it appropriately.

Each mushroom strain has a lot of different properties as they're from various pieces of the world. Some are a lot simpler to develop, some are significantly more powerful and some produce bigger yields than others. They additionally develop in different shapes, which you'll see once they open up. The accompanying photograph is of two unique strains (Pan-American and B+), which were set to reproduce on precisely the same day. The Pan-American shrooms have barely delivered any mycelium yet bigger mushrooms a lot quicker. In contrast, the B+ has made a layer of mycelium over the whole surface and is starting to top off with little mushrooms.

If you've never developed Psilocybin mushrooms, we suggest going with a strain called **Mexican**. This specific strain adjusts much better to temperature and humidity, so it, despite everything, has a battling chance regardless of whether you don't give it the specific right parameters.

Lighting

One of the most significant variables when growing Psilocybin mushrooms is the measure of light that they get and the nature of that light – they ought never to be given direct light. They can be developed utilizing daylight or a typical white bulb yet never legitimately on the substrate. To the extent the measure of light required, this can be dubious as mushrooms commonly develop on the ground in enormous backwoods, investing heaps of the energy in obscurity. If you cultivate them utilizing sunlight, you should simply leave your drapes open and spot the mushrooms aside of the window, ensuring that they're never in direct daylight. In case you're utilizing light in your home, you'll have to situate it with the goal that the light doesn't legitimately sparkle on the substrate.

Humidity

Humidity is necessary as it enacts the mycelium, which is the place the shrooms sprouts from. To give the perfect measure of humidity, you'll have to utilize a little nursery propagator. Hydrate the substrate utilizing packaged, assimilation or distilled water – never use faucet water. At the point when you include water, the substrate will start expanding, so you'll have to go gradually to ensure that it's equitably wet. When it is thoroughly doused, expel any extra water – if there's any water left at the base of the compartment, it might cause fungi.

Our mushroom units accompany a sack that you can use as a propagator. Although, if you need the ideal outcomes, we suggest getting an appropriate nursery propagator. To initiate the mycelium, you'll have to keep the humidity at a consistent 90% for in any event two days inside the propagator. It implies you'll have to add some additional water to the base of your propagator – not the compartment with the substrate. Ensure this water is additionally assimilation, distilled or packaged. Another approach to do this is to keep the cover on the holders, which concentrates humidity significantly more. After the initial two days, you'll have to bring humidity to around 70%, which is anything but difficult to do by tinkering with the little windows on your propagator.

Temperature

Temperature is another unimaginably significant parameter; mushrooms, for the most part, flourish somewhere in the range of **21** and **24°C**. So, if you need to deliver whatever number shrooms as would be prudent, we suggest keeping the propagator directly in the center. If it's cool where you anticipate growing them, you can generally get a warmed propagator or a warmed cover underneath your propagator. If you cultivate your mushrooms under **21°C**, they'll develop many increasingly slow, fewer shrooms – when the mycelium is dynamic and it just has a specific measure of time to create mushrooms.

Hygiene

To wrap things up, hygiene and a spotless domain are central to growing Psilocybin mushrooms. They're very delicate and they need a perfect and sterilized condition – never contact them with your hands under any circumstances. Make a point to utilize latex gloves when taking care of the unit at unprecedented and, if conceivable, a facemask. You have to abstain from modifying the environment around them, however much as could be expected – don't smoke, use antiperspirant or some other sort of shower item in the room that they're in or they may get debased and not develop appropriately.

Give your Psilocybin mushroom units the correct lighting, humidity, temperature and clean condition and they'll create a lot of hallucinogenic heads. **Growbarato.net** has all that you have to develop your one of a kind Psilocybin mushrooms, from the packs themselves to items, for example, warmed mats and thermo-hygrometers, just as full mycological study units.

Harvesting Hacks for Psilocybin Mushrooms

After anyplace somewhere in the range of 7 and 14 days, you should begin seeing the initial, not many mushrooms show up. From that point onwards, they'll fire springing up all finished – if you keep an eye on them a couple of times each day, you'll likely notification them getting bigger, growing a couple of centimeters daily. They may be prepared after around 3-4 days and in the wake of harvesting, you'll have to let them dry for a couple of more days before choosing to contemplate them.

While evacuating the mushrooms, put on specific gloves and squeeze them between your fingers, contorting somewhat – they should jump out straight away. They're very touchy and any place you contact them, they'll start to turn a dim shading for several hours. This is typical, don't stress over tainting or decay.

When you evacuate the entirety of the mushrooms, don't discard the holder – mycelium can remain dynamic for some time longer, contingent upon the conditions given. You may have a different yield of mushrooms after only a few days. If you've given completely flawless conditions, you may even get three or four adequate leaves, only one compartment.

Making Spore Syringes

If you have a spore print and access to syringes, you can without much of a stretch, make your spore-water syringe to use in the PF strategy.

Materials:

Spore, Syringe and needle, sterile water in a cup fixed with foil (15 mL for each syringe), Jet-Dry or comparable dishwashing wash operator, Alcohol light, Inoculation circle

1. Spot 25 mL water for each syringe you will get ready in a **1/2-16 ounces** jar fitted with a channel plate and cover, alongside two drops of dishwasher wash operator. The flush specialist can be found in most supermarkets. Search for a brand without vinegar or included fragrances. Its motivation here is to keep spores from amassing and adhering to the sides of the jar and 92 I PF Tek Improved syringe.

2. Seal and sterilize for 30 minutes at **15 psi**. Permit to cool before utilizing.

3. This technique is best acted in a glove box or before a stream hood to ensure sterility. Wipe your work region dwell with scouring liquor or **3%** hydrogen peroxide and permit to dry.

4. Light your liquor light and warmth the inoculation circle until intensely hot.

5. Unscrew the top, lift it somewhat from the jar, cool the circle in the sterile water.

6. Supplant the top freely on the cup and afterward go through the circle to pick a few spores from the print.

7. Lift the cover, twirl the circle in the water and afterward supplant the top. Rehash more than once for good measure. Distinct spores are minute; you won't have the option to see them in the water.

8. Bring some spore suspension into the syringe, delicately filling and exhausting the syringe a few times to protect an even conveyance of spores.

9. Supplant the needle spread, mark the syringe with the strain and date and spot it in a clean Ziploc sack until required. Syringes arranged like this can be put away in the cooler for a month or two. However, they are ideal whenever utilized as quickly as time permits.

Agar Spore Germination

This strategy is the same as the one utilized when making spore-water syringes, then again, actually, here the spores are moved to sans peroxide agar plates rather than water.

Materials:

Spore print without peroxide agar, Petri dishes, Inoculating circle, Alcohol light, Parafilm

1. In your glove box or stream hood, heat the inoculation circle in the liquor light until it sparkles super-hot.

2. Lifting the cover of the first Petri dish with your other hand, press the tip of the circle to the focal point of the agar to cool it (this likewise puts a thin film of agar on the circle, which will assist the spores with adhering to it).

3. Spread the plate and afterward go through the circle to pick a modest quantity of spores from your print.

4. Streak these over the Petri dish in an S-molded movement and afterward close the plate.

5. Re-sterilize and cool the circle before marking each plate.

6. After the shot, wrap the edges of each plate with Parafilm, mark them with any valuable data and brood agar side up.

Cardboard Disk Spore Germination

Materials:

Spore print, cardboard circles, J/2-16 ounces jar and top, Screw-topped test cylinders or vials (2 or 3 for each spore print), Malt-yeast agar arrangement (1 tsp. malt and a minuscule spot of yeast separate in 1 00 mL water), Pipette or eyedropper Tweezers, Alcohol light, Parafilm

1. Spot cardboard circles in J/2-16 ounces jar, alongside 1-2ml water and seal. Spot 5-10 drops of malt arrangement into test tubes and daintily seal. Sterilize the jar and cylinders for 15 minutes at 45 psi and permit to cool totally.

2. Spot all apparatuses and materials in your glove box or stream hood.

3. Warmth the tweezers in the liquor light until hot and permit to cool.

4. Open the jar and utilize the tweezers to expel one plate. Spread jar.

5. Gently contact the edge of the circle to part of the spore print. You ought to have the option to see the dark spores holding fast to the ring.

6. Open a test cylinder and drop the plate onto the base of the cylinder.

7. Rehash 3-5 times for every cylinder.

8. Make two containers of plates at any rate for every spore print.

9. Seal the cylinders with Parafilm and brood.

10. At the point, when the spores have developed and the circles are fully colonized, move a couple to singular peroxide-containing agar plates.

Chapter Nine

PF-TEK

The PF-Tek was created and first made open in 1992. The growing technique depicted depends on the PF-Tek yet incorporates a couple of changes that are reliable and areas I would see it better than the first PF-Tek. The PF-Tek for Simple Minds is as essential and as idiot-proof as it gets, yet isn't secure obviously. Tailing it will give you high odds of succeeding and a smart thought on the general procedure and course of events and set you up for higher-yielding Teks, such as utilizing entire grains and waste.

PF-Tek empowers cultivators to develop magic mushrooms without any preparation. This ensured technique for **Psilocybe Cubensis** cultivation is simple, modest and has a high achievement rate. We've seen that a ton of Magic Mushroom Shop clients has been requesting PF-Tek Instructions. We will manage you how to set up your PF-Tek substrate and how you have to make PF-Tek cakes. The PF-Tek for Simple Minds utilizes ½ half quart (240ml) canning jars or drinking glasses and a growing substrate made of vermiculite, earth-colored rice flour and water. The substrate is blended, filled in jars, sterilized and immunized with mushroom spores after the substrate is completely colonized. The substrate cakes are the natural product in a muggy compartment.

Neatness

By growing mushrooms inside on a nutritious substrate, you make conditions than favoring the development of the mushrooms, yet in addition to the development of countless different organisms (molds, microbes), a considerable lot of them possibly risky to the wellbeing. To guarantee that lone the ideal mushroom is developed, it is imperative to ensure neatness in the entirety of the cultivation.

Wash your hands with (antibacterial) cleanser and warm water before commencing work. A short time later, wipe them dry and rub with Lysol or isopropyl liquor (iso-propanol). Keep where you do the inoculation and fruiting residue-free and clean and don't acquire messy apparel or shoes. Individual hygiene is similarly significant. Dirty hands and even filthy hair are a hotbed for a wide range of undesirable microorganisms that can stop your cultivation venture.

Rudolf Materials Required for PF-TEK growing

Substrate:

- Brice flour
- Vermiculite
- Water

Supplies:

- Jars
- Covers
- Aluminum foil
- Pressure cooker or standard container
- Grow sack
- Spore syringe, spore vial or spore print
- Syringe
- Alcohol burner or Torch-lighter
- Face Mask/Respirator
- Gloves
- Mixing bowl
- Scale
- Measuring cup
- Fork/spoon
- Awl, nail and mallet, drill
- Marker
- Tape or names

Materials:

Most materials are effectively accessible at the nearby shops.

Vermiculite:

Vermiculite is produced using an ordinarily happening mineral - **mica**. Squashed mica containing water is warmed and extends to a volume a few times more noteworthy than that of the untreated mica. Vermiculite can hold a few times its load in the water and it gives the substrate a breezy structure. Vermiculite is accessible in a few evaluations; the center and the exceptional center evaluation are generally reasonable for cultivation purposes. By and large, you can get vermiculite in the garden and hydroponic stores, in certain areas likewise in pet shops.

Brown Rice flour -BRF:

BRF is accessible in wellbeing food shops either as of now ground. In some cases, however, there is just entire earth-colored rice available. For this situation, you can crush the rice either in the shop. If this alternative isn't accessible, pound you're utilizing an electric espresso processor. BRF is best saved cool and dry for delayed timeframes since it can without much of a stretch become rank due to the fat substance of its husk. If you can't discover BRF, you can likewise utilize entire rye flour, ground millet or ground millet-based birdseed with comparative outcomes.

Water:

The water utilized for the substrate readiness ought to have drinking water quality. Faucet water is ordinarily OK, however, in case you don't know about it, better use packaged drinking water or mineral water.

Spore syringe:

A plastic syringe with needle appended containing a **10cc-12cc** suspension of mushroom spores in water.

The shade of the suspension shifts from translucent to somewhat violet contingent upon the amount of the spores in the arrangement.

Jars:

The jars ought to have a substance of around ½ half quart (~240ml). You can utilize either canning jars or drinking glasses. The primary necessity is that they are tightened and without shoulders. So, you can slide the cake out of it in one piece once it's colonized. More magnificent jars take any longer to colonize and are not suggested.

Substrate arrangement:

For one ½ half quart jar (~240 ml) you will require:

- 140 ml of vermiculite
- 40 ml of brown rice flour
- Few vermiculites to fill the jar to the top (application. 20 ml)

Water

For six jars, this adds up to:

- 3.5 US cups of vermiculite
- 1 US cup of earth-colored rice flour

Note:

½pt (US half quart) = 1cp (US cup) = 236ml (milliliter) = 236cc (cubic centimeter) = 1/4 qtr. (US quart)

Put the necessary measure of vermiculite for all the jars of one bunch (for example, six jars: 6 x 140 ml = 840 ml ~ 3.5 US cups) in a bowl.

Pour water gradually over the vermiculite while blending with a spoon. Be mindful to just place that much water in as it very well may be consumed by the vermiculite. Mix it well, so all the vermiculite is consistently drenched with water. At the point when you tilt the bowl, you should see only a little water beginning to originate from the vermiculite.

It is the point at which the right water content is accomplished. If there is a lot of water in the bowl, pour the wet vermiculite in a sifter and let the abundance water channel for a moment. At that point, the vermiculite will be at the field limit, which is excellent.

Put the necessary measure of the **BRF** (for example, 6 x 40 ml = 240 ml = 1 US cup) into the wet vermiculite on the double and blend it in with the spoon. The objective is to consistently cover the wet vermiculite particles with a layer of BRF.

Fill the blend in jars ½ inch (1cm) under the top. It's essential to fill the substrate in the jars without tapping any stretch of the imagination. It should remain breezy and free to give ideal conditions to the development of mycelium. Be mindful so as not to leave any substrate on the upper edge of the jar. If you weren't sufficiently cautious and there are some substrate bits at the tip, take a spotless wet material and wipe the upper bit of the jar clean. In any case, contaminants can begin at those spots and work their way into the jar.

Top off the jar with dry vermiculite to the top. This layer prevents airborne contaminants from arriving at the hidden substrate if they figure out how to come in during the inoculation and incubation.

Take a 7in (14cm) wide stripe of aluminum foil and overlap it in the center. Put the foil above the opening of the jar, as appeared in the photos. In case you're utilizing jars with metal covers, you can jab four openings at the very edge of every ceiling with a little nail and mallet and screw the top on. The openings ought to be somewhat more significant than the measurement of the syringe needle.

Crease the foil edges up and press them together, so you get a decent aluminum foil top. At that point, take a bit of foil estimating **5in x 5in** in and put it over the initial two layers (separately the metal top in case you're utilizing tops), leaving the edges of the foil coming to down since it must be lifted again during the inoculation.

Sterilization

Pour around **1in (2.5cm)** of water into the pressure cooker, don't place more water else it will come into the jars and change their water content. At that point, stack the jars into the pressure cooker. The utilization of a rack to keep the jars from straightforwardly contacting the base of the cooker is firmly suggested. Put the cover on and carry the cooker to the necessary pressure (15 psi = 1atm over barometrical pressure) gradually over a time of **15 minutes** on a medium fire.

If you heat the cooker too quickly, this can make the jars split. When the steam starts to get away from the rocker or the vent at the highest point of the pressure cooker turn the warmth back, so just a little, consistent steam stream endures from the vent—starting here on, pressure cook for 45 minutes. Contingent upon the pressure cooker model, the cooking procedure works somewhat unique, so in case you're inexperienced with pressure cooking, counsel the instruction manual or somebody who utilized pressure cookers previously. After 50 min, take the cooker from the fire and let cool for in any event 5 hours or far and away superior short-term. If you never utilized a pressure cooker, look at this record about the right pressure cooker use.

If you can't discover or purchase a pressure cooker, you can likewise **sterilize** the jars utilizing the main container with a top.

For this situation, steam the jars for 1.5 hours in a pot cover on. Utilize just around 1 inch of water at the base. You may need to add some water to the pot during steaming because of dissipation.

Disinfecting Substrate in a Regular Dish

- Take a tall dish with a good fixing spread, where you can place all the jars in.
- Put a rack on the base of the skillet or use covers for this. It will keep contact from the jars with the hot base and keeps the jars from breaking due to the warmth.
- Open the front of the substrate jar marginally and spread with foil. Put every one of your jars in the container.
- Fill the skillet with water to a limit of 1 centimeter under the edge of the jars.
- Slowly heat the water to the breaking point.
- Leave it to steam for an hour and a half with the spread on the dish. Keep the fire as low as could reasonably be expected with the goal that the water keeps on steaming. Once in a while, ensure there is sufficient water in the container.
- Turn off the fire following at least an hour and a half of steaming. Steaming for a more extended period is additionally a choice to make sure of fruitful sterilization of the substrate.
- Leave the jars in the skillet to chill, this can take a whole night. Hold up at any rate of five hours.

- Take all the jars out of the skillet and mark each jar with a number or letter, so you can disclose to them separated.
- The jars are currently prepared for the inoculation with spores.

Disinfecting substrate in a pressure cooker

- Put a rack on the base of the pressure cooker or use covers. This will keep contact from the jars with the hot base and keeps the jars from breaking in light of the warmth.
- Fill the pressure cooker with about six centimeters of water. Or, on the other hand, adhere to the instructions of the pressure cooker's manual.
- Place all the jars on the rack and close the pressure cooker.
- Heat gradually and sit tight for it to arrive at the correct pressure, i.e., 15 psi or one air.
- Let it steam for a 45-an hour and afterward turn off the fire.
- Leave the jars in the container to chill, this can take a whole night. Hold up at any rate of five hours.
- Take all the jars out of the dish and name each jar with a number or letter, so you can disclose to them separated.
- The jars are currently prepared for the inoculation with spores.

Tip: Smell the substrate while disinfecting. You will comprehend what it smells like and you can later smell whether the smell has changed. A harsh and solid aroma can show a tainting.

Tip: Keep the jars for a limit of multi-week on room temperature in a dim and without draft place. Check the jars after this period for defilements to forestall immunizing the polluted jars with spores. Dispose of the substrate of defiled jars promptly and sanitize the region and jar.

Inoculation

It is the second piece of the PF-Tek technique. You should work sterile and cautiously. Ensure you completely clean the territory you work in before infusing the substrate jars with spores. The utilization of purifying gel for your hands and a respirator is suggested. You can likewise make a **'glove box'** or inoculation space, yet it isn't essential if you continue spotless and sterile. When immunizing, you include the spores through a syringe to the PF-Tek substrate you made before. The spores create to mycelium that will gradually colonize the substrate of the jar. Expect a 1ml spore answer for one jar. You can likewise utilize increasingly (2 ml). At that point, the jar will colonize quicker.

Procedure

When the cooker is cold, place them on a spotless surface, have a liquor light or a lighter and the spore syringe prepared. Shake the spore syringe to separate the spore clusters.

It's vital that there is a little air pocket of air in the syringe to shake. If this isn't the situation, at that point, you can suck roughly **1cc** of sterile air into the syringe by setting the tip of the needle into the fire and gradually pulling the unclogged back.

Extricate the foil from the entirety of the jars so it very well may be lifted effectively on vaccination.

Take the spread from the needle and warmth it over the fire until gleaming red. Let cool for a couple of moments.

Take the upper foil layer off and set aside.

Penetrate the foil at the edge of the jar with the needle application. 1in (2.5cm) profound and infuse the spore suspension towards the inward jar surface. You should see a little let running fall the internal surface of the jar towards the base. Each jar is vaccinated on four similarly separated focuses. You should utilize **1 - 1.5 ml** of the spore suspension per jar, so one 10ml syringe is adequate for 6 - 10 jars.

Put the foil on once more. Fire sterilize the needle again after immunizing three jars to forestall cross-contamination just if a jar wasn't appropriately sterilized.

When the jars are immunized, overlap the foil edges up and press them solidly together, so you get a decent aluminum foil top. Compose the inoculation date and the species/strain data on the foil with an all-surface felt tip pen. If you contact something other with the needle during the inoculation procedure aside from the foil surface of the base foil layer promptly fire sterilize the tip once more.

How would you infuse the substrate jars with spores?

- Check your substrate jars for pollutions.
- Disinfect your working environment, wash your hands and put the respirator on. Gloves are discretionary.
- Prepare all that you need: Spore syringe, burner, jars and foil
- Prepare the spore syringe.
- Remove the foil from the spread. If you just utilized foil, expel only the top layer of foil.
- Stick the needle of the spore syringe through the spread.
- Move the needle out of the jar and squeeze the sprayer. This permits the spores to spread over the edge of the substrate jar.
- Heat the needle super-hot after each infusion yet hold up till it's chilled off adequately before infusing it once more.
- Inject the jar on four places close to the edge and conceivably in the center.
- 1 ml of spores' arrangement per jar is adequate (if you utilized 240 ml jars).
- After inoculation, spread the substrate jar with foil.
- Repeat stage 5 to 13 for the PF-Tek substrate jars.

Incubation

The jars ought to be put away at 21-27°C (70-81°F), the hotter and the better, yet not surpassing 27°C. You can also construct an incubator to suit the jars.

Incubator: The immunized jars grow quickest if they are put away at a temperature of 27°C (80°F). The best incubation temperature for P. Cubensis would be 86°F, however since the jars themselves are a couple of degrees hotter than the environmental factors (mycelium radiates heat when growing), 80°F is a decent and safe incubator temperature.

You can fabricate a successful incubator by utilizing two plastic boxes of a similar size and an aquarium radiator. There are a few kinds of aquarium radiators. At the point when you're purchasing a radiator, ensure that it is of the "completely submarine" type.

Connect the radiator to the base of the principal enclose and pour as much 27°C warm water that the warmer is lowered. Alter the warmers' indoor regulator, so the radiator just closes itself off at 27°C.

Put some spacers on the base of the container; they convey the subsequent box and keep it from contacting the radiator. In the above picture, four jars are utilized. You could likewise utilize blocks, stones or something comparable.

Following a couple of hours, measure the temperature again and change the radiator if vital, so the water temperature is **27°C**. At the point when the container is vacant, it will glide on the water. 2/3 the stature of the container, assuming the upper box is set up stacked with jars and laying on the spacers.

You can now place the immunized jars into the crate. Spread the jars with a cover to conserve the warmth and to keep the jars dim.

Note: The water level drops in certain weeks by vanishing. Hence you need to fill some new water every once in a while to keep the water level sufficiently high. Never let dissipate so much water that the warmer isn't lowered in water any longer.

Keeping the jars warm, you should see the principal indication of germination following 3-5 days as splendid white bits. It is mycelium. If anything develops that isn't white, for example, green, dark or pink, at that point, the jars are tainted and their substance must be disposed of and your perfect procedures need some improvement. After the jars are discharged and the jar is washed with cleanser and high temp water, it tends to be utilized once more.

Contingent upon the temperature and the practicality of the spore syringe, it takes 14-28 days for the mycelium to colonize the entire jar. When colonized, store the jars at ordinary room temperature, about 21°C (70°F) to start sticking.

Try not to open the jars to coordinate daylight. Backhanded daylight (= the common light that lights up a room because at daybreak) or a low wattage light (cool white fluorescent light is perfect, radiant light is less reasonable) for 4-12 hours daily is adequate.

Inside 5-10 days (with certain mushroom strains, it can anyway take as long as 30 days) pinhead-size collections of mycelium should shape. These purported pins speak to the start of mushroom development. In the next few days, add little mushrooms with earth-colored heads become noticeable. At the point when this is the situation, it's an ideal opportunity to birth the cake into the fruiting compartment where the mushrooms can create to develop.

A few strains don't effectively create pins. For this situation, but the colonized jar enveloped by a plastic sack in the ice chest expedite and afterward continue to fruiting the following day, regardless of whether the cake doesn't show sticks yet. This virus stunning, for the most part, helps trigger sticking to some degree.

Notes:

The whole colonization of the PF-Tek substrate can take as long as about a month, contingent upon the conditions. Following three to seven days, you will see the principal indications of the framing of the mycelium. You can perceive this by the white stains of strings that develop from the spots where you infused the spores. At the point when the whole substance of the jar is white because of the mycelium, it is ideal for holding up one more week so that within the substrate is completely colonized also.

Check the substrate jars during the incubation time frame for pollutions. A tainting can be begun by the distinctive shade of the growing mycelium. The mycelium is white; defilements can be velvet, green, orange, dark, splendid yellow or dim. If you find pollution in a jar, it is necessary to discard the substrate right away. Wash your hands altogether a short time later and sanitize the spot and the immediate condition of the jar. Take notes on which jar got defiled, after how long it was evident and how the tainting looks.

At the point when the mycelium has colonized the whole PF-Tek substrate, the possibility of contamination is extremely little. The mycelium, which is a living life form, will battle numerous pollutions without anyone else and attempts to endure!

Fruiting

The fruiting of the cakes can be cultivated in any kind of compartment that can be approximately fixed and has at any rate one translucent side, ideally on the top. Appropriate compartments are a plastic can, Rubbermaid holder, terrarium and aquarium.

Put a **1/2 inch** layer of moistened perlite (pdf) or extended dirt pellets or even a wet paper towel at the base of the compartment and birth the cakes onto this layer by letting them slide from the jar.

Then again, one would first be able to apply a packaging layer. Some of the time, the cake doesn't slide out of the jar effectively without anyone else. You simply need to flip around the colonized jar in your grasp and pummel the hand delicately against the palm of the other hand. All of this will make the cake slide against the top and it tends to be birthed easily if you have a greater fruiting chamber (a greater plastic holder or a terrarium) you can place in more than one cake to an organic product.

The separation between the cakes ought to be at any rate 2in (5cm) for the mushrooms to have space to develop. Take this sheet off once per day and fan the let some circulation into with a bit of cardboard. If the base layer starts to dry out, shower it with some water to keep it clammy since this layer gives dampness to the air to remain muggy. Try not to splash the cakes straightforwardly.

Handle the cakes as meager as could be expected under the circumstances. However, when you do, it generally washes your hands altogether. Over the following 7-14 days, the cakes will start to stick (if they haven't started to stick in the jars yet) and the little mushrooms will become enormous in only 2-5 days and as soon as the tops open they can be harvested.

This synchronous development of all mushrooms is known as a **flush**. After the mushrooms have become large, there are generally a couple of little, hindered mushrooms left finished;

they are called **prematurely ends**. They can be perceived by their blackish heads and the way that they quit growing sooner or later. Still, they are acceptable to utilize, except if they are spoiled.

It's pivotal that you harvest all mushrooms. Likewise, the prematurely ends, after the flush. This is most handily cultivated if you harvest the mushrooms off by tenderly winding and detaching them the cake with clean hands. Alternatively, you can dunk the cakes after each flush and it can build the flush size essentially.

After roughly multi-week little mushrooms start to shape again and develop during the following days, this cycle can rehash itself up to multiple times once in a while considerably more. After that, the cake is depleted; it creates no more mushrooms and can be disposed of. They can be likewise used to begin outside beds. Some of the time, green form assaults the cakes even before they are depleted. If so, expel and dispose of the sullied cakes promptly to forestall the spreading of the tainting.

The Cultivation

Fundamental: Colonized substrate jars, fork/spoon, spread, cultivation pack, paper cut, plant sprayer, gloves, disinfectant arrangement, respirator, vermiculite and perlite are discretionary

If the PF-Tek substrate has completely developed following a month and doesn't contain any contaminations, it's the ideal opportunity for the last piece of the PF-Tek strategy, i.e., setting up the substrate cakes for the growing of mushrooms. You have to remove the cakes from the jar first to develop the mushrooms from the substrate.

- Open the front of the PF-Tek substrate or expel the foil.
- Carefully expel the vermiculite on the PF-Tek cake utilizing a sterile fork. You may discover mycelium here and you can expel this along with the vermiculite.
- Take a spread or plate that is somewhat bigger than the substrate jar and put it on the jar.
- Place the jar and cautiously tap the substrate-free.
- Let the substrate float out of the jar and on to the spread or plate.
- Put your PF-Tek cake on the spread in the cultivation sack. No doubt, there will be space for two cakes in a single cultivation sack.
- Spray some water in the cultivation sack utilizing the plant sprayer to get the humidity to the right level.
- Repeat stage 1 to 7 for all your PF-Tek cakes.

Best conditions for mushroom development:

- The optimal temperature for your mushroom cultivation is 24 degrees Celsius.
- Maintain the humidity in the mushroom cultivation sack on 95%.

- Refresh the air before shutting it once more. An adequate measure of oxygen and not all that much carbon dioxide.
- An adequate measure of light, however no immediate daylight.

After one to a little while, you will have the option to harvest your mushrooms from your handcrafted PF-Tek development unit!

After the initial flush:

Harvest each mushroom. Indeed, even the littlest. After the main flush, you should drench the substrate cakes again for the following flush. Along these lines, the substrate will have adequate dampness for another flush of mushrooms.

1. Take a perfect container or bowl with clean drinking water.
2. Place your PF-Tek cake in the water and ensure it remains submerged.
3. Let the PF-Tek cake drench for twelve hours.
4. Remove the PF-Tek cake from the water and set it back in the cultivation pack.
5. You can harvest about four flushes of mushrooms from the equivalent PF-Tek cake.
6. After each harvest, in the wake of expelling the mushrooms, you can rehash stage 1 to 5.

Dunk and Roll

Another strategy is to quickly drench the cake from the jar and move it through vermiculite. Some mycologists show signs of improvement results along these lines. The vermiculite causes the dampness to arrange all the more step by step.

1. Take a perfect pail with clean drinking water.
2. Put your PF-Tek cakes in the water and ensure they remain under.
3. Let the PF-Tek cake drench for twelve hours.
4. Take a perfect plate and sprinkle it with vermiculite.
5. Take the PF-Tek cake out of the water and move it through the vermiculite.
6. Repeat stages 1 to 5 with each cake.
7. Only roll the cake through vermiculite while planning for the principal flush. After the main flush, a dousing of the PF-Tek cake will be adequate for the following flush of mushrooms. Rolling won't be essential at that point.

Tip: If you are utilizing different PF-Tek jars, you can try different things with the two methods. Utilize the **'Dunk and Roll'** technique for half of your jars. It permits you to look at and discover which strategy works best for you.

Chapter Ten

Flat Cake Tek

Supplies:

- Casserole dish (glass just) with plastic cover (Pyrex brand)
- Saran wrap
- Vermiculite
- 3 completely colonized PF cake fourteen days after 100%colonization.
- Needle
- Distilled water
- Fork

Directions:

1. First, sterilize your territory with Lysol disinfectant splashed noticeably all around. At that point, use scouring liquor to sterilize your table. At last, use scouring liquor to clean yourself and any devices you are working on.

2. Pre-sterilize vermiculite, for a 25-an hour at 325 degrees.

3. Now birth the jars. First, scratch of the verm layer and entail three jars for every level square goulash dish.

4. Expel any early-stage or pins that have begun growing on the cakes. Presently separate cakes into little pea-size pieces. You almost disintegrate the cakes.

5. Next, you blend restrained water with your vermiculite. Permit the verm to be soaked, not wet. Possibly just a drop comes out between your fingers when a bunch is crushed.

6. Utilize the fork to pack the mycelia equitably. Ensure it is as even as conceivable this will help later on.

7. Shortly, spread the verm. Over the cakes equally, covering the mycelia altogether. Utilize your hand to search the vermiculite with the goal that it is overall quite cozy.

8. Utilize the plastic wrap over the goulash dish and jab 10 or 12 openings arbitrarily in the cling wrap. At that point, name your plate. This picture shows the plate secured with plastic wrap, verm and marked and prepared for incubation.

9. At long last brood, your holders in all-out dimness for three to five days until the **mycelia** is thick and appears as though an altogether colonized PF jar once more. The pieces that you separated will become together and appear as though one cake now.

10. Try not to permit light in the incubation room and don't utilize warming cushions.

11. When the dishes have become through, birth them by scratching the top verm layer off and afterward utilize a fork cautiously to pull the cake away from the edge of the meal dish. Presently turn the cake over on the cling wrap and the top to the goulash dish.

Tip: While setting the plate inside the holder, a decent indication is to take the piece of the cling wrap hanging over the meal dish top and take advantage of your terrarium divider. It helps the mushrooms colossally.

CHAPTER ELEVEN

Rye Grain Tek

It is a straightforward grain Tek. I like to utilize rye because of the way that I case the grain straightforwardly and don't use it to bring forth, which I discover better. The rye holds more water, so it considers bigger blooms others may contrast. We, as a whole, have our specific manners of doing things. This is completely what I do.

1. First, I set up the grain by placing it in a sifter and shaking the residue off of it each cup in turn. At that point, put some grain or rye for my situation into a quart jar and fill the jar with water and let it drench for 24 hours. My purpose behind doing this is microbes endospores, that structure on grain can endure the PC. So we let the little fuckers grow and afterward PC them.

2. Take the rye and spill out the water. You're going to see that it has turned piss yellow. At that point, take the grain and put it into a pot. Include water, so the grain is pleasantly secured and stew it for 40 to 45 minutes. You'll see a portion of the rye grain beginning to blast and the grain has turned earth-colored and swollen up a considerable amount.

3. Spill the water out and put the grain into a sifter and let it channel. I like to flush it off after the stew.

4. Presently load the rye into the jars, I like quarts, so that is the thing that I use. Leave sufficient space at the highest point of the jar for you to shake the grain around. I didn't have any jars with just grain, yet you get the thought. No additional water should have been added to the jars, only the grain. I like to set up my jars like the metal cover and afterward the band.

I like this Tek for a couple of reasons. Ideal for composing on and simple to make with a cutout in an hour, you can make enough plates forever. At that point, PC the jars at **15psi** for 1 hour and let them cool. A little clue, if you take them out when they are as yet hot and shake them, they won't cluster up when they cool.

5. When cooled, I like to include somewhere in the range of 2 and 3 mil to each jar. Two if I made the syringe, three if I got one to add a strain to the assortment. I push the needle directly through the Tek over the little gap in the metal top. Simple to discover the gap if you hold the jar inclined toward a light. I spread with a bandage since they come sterile as of now.

6. When the jars are wholly shot up, stake them like a crazed nut. Blend them up as well as could reasonably be expected, so the spores start from the very beginning of the jar. In 9 days to 3 weeks max, your jars will develop.

7. Make your packaging with whatever packaging blend you choose to utilize. Let them sit in the incubator for 4 to 7 days at that origin.

The regular all-out time from the time I shoot them up till the major harvest with this grain Tek is 4 to 5 weeks.

Chapter Twelve

Popcorn Tek

Popcorn Tek isn't as famous as different methods. Certain cultivators, particularly tenderfoots, accept that shrooms don't colonize as pleasantly on popcorn as they do on rye. Then again, individuals who have culminated in this strategy depend on their adequacy and straightforwardness. Popcorn Tek is a tranquil, modest and open strategy for growing grain bring forth for various types of magic mushrooms. This guide is a bit by bit process for tenderfoots who need to develop their magic mushrooms utilizing popcorn as grain brings forth.

Popcorn Tek Supplies and Equipment:

Try not to be confounded. The items recorded here are for the inoculation stage as it were. More things will be included in different advances.

- Pressure cooker
- Wide-mouthed jars, 1 quart each
- Scotch tape

- Polyfill – this is stuffing inside pads
- Latex gloves
- Rubbing liquor
- Strainer
- Aluminum foil
- Hand sanitizer
- Lighter
- Lysol (air sanitizer and deodorizer)
- Spore syringes with spores from your preferred example
- Alcohol light or Bunsen burner

Making Grain Spawn Out of Popcorn

After setting up the supplies, the primary activity is to drench two packs of popcorn. Put two packs of uncooked popcorn in a primary bowl and top it off with water until it arrives at a stature of two crawls over the top degree of the popcorn. Permit the uncooked popcorn to drench for an entire 24 hours. These two sacks of popcorn will be sufficient to fill in any event six jars.

While drenching your popcorn, set up the jars for the succeeding advances. Stick two gaps into the covers – a huge one in the middle, followed by a little one towards the edge. It could be possible via cautiously pounding a nail or penetrating them with a screwdriver.

The enormous gap in the inside is expected for Polyfill. It will permit gas trade as CO_2 gets away from the jar without giving contaminants access. The littler gap is for the spore syringe.

Stuff the Polyfill in the middle gap and put a bit of scotch tape over the littler, inoculation opening. A few people incline toward utilizing micropore tape over this gap to make it increasingly permeable.

In the wake of letting the popcorn drench for 24 hours, dump them into a sifter and flush them off rapidly. Put the doused popcorn into a pot loaded up with new water to around two creeps over the highest point of the popcorn. Put the revealed pot on the oven, turn on the oven and heat the water to the point of boiling.

Ensure it remains at a low bubble for 40 minutes. Mix occasionally to forestall any popcorn at the base of the pot from consuming.

After 40 minutes, check the several corns. You'd realize they're prepared if they can be handily crushed between your fingers. When the corns have accomplished this delicate quality and surface, turn off the oven and strain the hot popcorn. Blend them around for 15 to 20 minutes until all water is depleted and the corns are dry to the touch. The objective here is to keep overabundance water from getting into the jars.

Fill the jar to 2/3 to the edge. At that point, close the cover. When all the jars are prepared, spread them with rock-solid aluminum foil.

Spot the jars inside the pressure cooker and start cooking. Hold up until the pressure measure gets up to **15psi** and begin the clock for **50 minutes**. After the dispensed time, turn off the warmth and permit the jars to chill for the time being.

Inoculating Popcorn Grain Spawn:

Set up the work territory while the jars are cooling. Splash the zone with Lysol to kill any microbes. After the jars are cooled down, you can immunize the popcorn. You will require the accompanying supplies:

- Jars loaded up with popcorn
- Scotch tape
- Lighter
- Folded up paper towel
- Spore syringe with the tops still on them
- Gloves
- Hand sanitizer

Be aware of neatness and sterility starting here onwards to maintain a strategic distance from contamination, which will destroy the shrooms. Wipe surfaces utilizing paper towels wet with liquor. Ensure to clean your hands with sanitizer and put on the gloves.

Get the syringe and shake it hard for a moment to get the spores to spread out in the syringe and to break separated any spore bunches. Light a fire and warmth the needle from the syringe by setting it near the fire until it gets intensely hot. Let the needle cool until it is not at the point red. Now, cut the scotch tape in your first jar. Point the needle towards the jar's inner side. Infuse just **2mL** of fluid from the spore syringe.

Expel the syringe and spread the opening. You simply punctured with another layer of scotch tape. Put that jar in a safe spot and rehash the procedure until you have vaccinated all the jars or until you have expended your spores.

After finishing all the jars, place them in a wardrobe where they can avoid sun introduction and advantage from a low-temperature condition. Keep the temperature around **73 degrees** Fahrenheit.

Mycelium should begin framing in 5 to 14 days. Watch their development and advancement until the jars are **70%** loaded up with mycelium. Shake them side to side to permit the mycelium to spread all through the jar. In the wake of colonizing the whole container, hold up five additional days just to be sure that the corns in the center are also inhabited.

Inoculating Substrate with Popcorn Grain Spawn:

You can pick any substrate which you find advantageous and viable. Probably the most well-on substrate decisions are sawdust, wood chips and rye as long as they are purified. Accepting you have numerous jars secured with mycelium, you can discharge them into a huge plastic tub loaded up with your picked substrate. Shake the corns around just to blend them with the substrate.

Following ten days, the tub will be loaded up with mycelium. Present outside air once in a while by forgetting about the tubs in a room that has some natural light in it three times each day for 30 minutes each. The objective is to present outside air and dispose of the CO_2. Try not to stress a lot over pollution because now, the mycelium is no longer as delicate.

Following half a month, you will see the shrooms shaping inside the receptacle. They'll begin as pins yet will later cultivate into enormous, grown-up magic mushrooms.

CHAPTER THIRTEEN

Fast Food of the Gods Method

Get a Rubbermaid pail molded holder that will fit in your microwave. Each progression goes on in this one holder. In the base, put two cups of vermiculite. Utilize a spatula to blend in enough refined water to cause the vermiculite about as saturated as it can be without feeling soaked (for the most part about a cup). The following dry ingredients can be included, each in turn or combined. The thought is to cover the wet vermiculite particles with the dry powder as you mix the blend in with the spatula. It sounds trifling. However, it has a significant effect.

Ingredients:

- 1/4 cup earth-colored rice flour
- 1/2 teaspoon dextrose
- 500 mg glycine
- 1/2 teaspoon shellfish shell powder
- 1/2 teaspoon follow minerals (gypsum powder may work)

Where do you get this stuff? - All are accessible at wellbeing food stores. Dextrose is additionally available from wine-making/lager preparing stores or diabetic gracefully organizations.

After making the blend, softly pack it and spread this layer with **"1/2 to 1"** dry vermiculite. Microwave the basin for 8 minutes with the top somewhat off. Allow cooling - in the microwave. (If you take it out and put the top on close, the top will get sucked in.) Now you're prepared to immunize.

I favor inoculation with mycelium water. However, many have pushed spore water; one may work. **Mycelium water** yet is a lot quicker and has less possibility of contamination. Injection around the edges and a few spurts in the center (5-15cc) will get things moving in a rush.

Wrap the aluminum foil outside of the compartment to the degree of the highest point of the vermiculite. Set it on the shelf and overlook it. Organic products will show up in the container in around three weeks (at 75 F) after the subsequent flush spurt in another **50cc** or so of water. At some point, these cans flush for a considerable length of time. When it looks old and crapped, drop-in a sterilized dairy animal's patty and more water. Again you may get more flushes.

Fire Warning - Cultivators ought to be mindful of screening their microwaving with the goal that the substrate doesn't dry out and of being cautious the first run through utilizing another microwave.

Chapter Fourteen

Psilly Simon's Method

The entire procedure is a mix of the rice cake and the Oss and Oeric methods and takes around a month and a half. The primary distinction from this technique is that the spores are dropped legitimately on the rye medium as opposed to developing them on agar first. The explanation behind the agar step is to expand sterility and guarantee just one strain of dikaryotic mycelium pervades the rye. Nevertheless, with the immediate spore technique, numerous strains are compelled to battle it out in the rye permitting the most grounded to rule the jar and natural product. I have had no sterility issues with rapid spore inoculation up to a moderately clean business spore print is utilized. Most are basic approaches to keep things clean without building a sterile box.

Equipment:

- **Pressure canner:** Fit for supporting 15 pounds. Size doesn't make a difference as long as you can do all the jars in the long run.

- **12 wide-mouthed, quart measured canning jars:** During the canning season, these can be found at any market. In the slow time of year, they are more earnestly to discover.

- **Spore print:** FS Books have great prints. Check High Times for the location.

- **1200 ml real grain rye:** Not creature feed! Get it at a wellbeing food store. Rye is better than rice since rice gums up the sides of the jars so you can't perceive what's growing inside it.

- **One sack planting soil:** Peat greenery/pearlite/vermiculite blend just (no earth)

- **Styrofoam cooler:** Sufficiently huge to hold all the jars

- **Transparent/translucent plastic board:** You can either utilize Plexiglas or great home improvement shops to convey boards to cover fluorescent lighting units. These can be cut with scissors without any problem.

- Lysol shower

- Sandwich measured Zip Lock Bags

- Antibacterial cleanser

- Heavy obligation zircon encrusted tweezers

- Scraper

- **Flame source:** Lighter, liquor light and so forth

- Saran wrap and tin foil

- 1 gallon Distilled water

- Water shower bottle

1) Wash the jars with an antibacterial cleanser. Utilize a dishwasher as well if you have it. It isn't urgent to get profound clean right now, simply be slick.

2) Include 100 ml rye and 175ml purified water to 3 canning jars. Close the jar with the cover. The covers will remain that way and all through their utilization. Keep the vault free yet secure.

3) In a spotless compartment, blend some dirt in with pure water until it is elastic to the touch and doesn't leak any water. The dirt ought to be wet yet not fluid. You need "sodden soil," not mud. Blend enough of the dirt to fill a canning jar freely. Try not to pack the dirt in, simply drop it in the jar till it's full. Screw the cover on freely, however safely.

4) Place the three rye jars and the dirt jar into the canner. If you can fit multiple containers in there proceed, it will spare you time. Simply make sure to get ready one jar of soil for every 3-4 jars of rye. Follow the directions the canner to sterilize the jars at 15 pounds for one entire hour. If you can, it is a smart thought to let the steam develop a piece before shutting the valve. It isn't necessary to utilize purified water in the canner.

5) When it is set, let the canner cool to room temperature. When it is sheltered to deal with, you can expel the jars and let them cool independently. Jars must be cooled to room temperature before proceeding. Store the dirt jar some spotless and fix the cover. Softly shake the rye jars to release the rye.

6) Repeat stages 2-5 until all the containers are wrapped up. You ought to have one jar of sterile soil for every 3-4 jars of sterile rye.

7) Here's the precarious part. Many people gripe about contamination. However, if there is a great strategy to vaccinate the jars, you won't think that it's an issue. I utilized the procedure with my last cluster of 12 jars and **NONE** of them were debased! Try to open the jar tops as meager as workable for a short timeframe as could reasonably be expected. Additionally, make an effort not to remain over the jars. They are quickly aired out.

Take a shower. Wipe off a work area or table and wipe it with an antibacterial cleanser. Shower it with Lysol. Screw off the vaults yet leave the tops on. Clean your hands again with an antibacterial cleanser.

Ready the spore print. Try not to remove it from the pack. Fire the scrubber and the tweezers until they sparkle. Let them cool at that point. The tweezers are utilized to hold and open the sack while the scrubber gathers spores. The spore print never leaves its sack. Try not to shower Lysol close to open blazes.

When a noticeable bunch of spores has been scratched off rapidly, convey them to the scrubber on a jar. You can see tremendous spores. It doesn't take a lot. Air out the jar only enough for the scrubber to enter and drop in the spores in. Close the cover and screw on the vault solidly. The cover ought to have just been aired out for around 2 seconds, insufficient to pollute it. When all the jars are vaccinated, shake them until all the rye is free and the spores are disseminated. Slacken the tops.

8) Place the rye jars in the Styrofoam cooler, close the top and pause. It takes around one to fourteen days for the mycelium (fluff) to saturate the jars. Little clusters of white fluff will show up in the jars. At the point when the development is about half, saturate the jars and let the fluff develop once more. The proportion of rye to water to be excessively dry and accept twice the length of **100 ml to 175ml**. Likewise, it utilizes less rye, so the jars are pervaded quicker. For the most part, It takes ten days. If anytime, you see any non-white fluff or non-rye gunk in the jar, then it is polluted. Dump it out. There's no desire for it and it isn't beneficial to ingest. It could be deadly or more terrible. Be coldblooded. That is the reason you completed **12 jars**, so you could forfeit a couple if necessary. Ordinary temperature is considerable for the entire development cycle, yet don't keep them next radiators or cooling. Albeit a few sources recommend keeping the **85 degrees (f)** room temperature is remarkable.

9) When all the jars are prepared, remove them from the cooler. Now you get 1.5-2 creeps of the sterile soil onto the rye.

It is called **packaging**. There are two different ways to do this without giving up sterility:

A: You can turn the cooler sideways, spread within and the opening with cling wrap and cut two gaps in the cling wrap over the opening to make a sterile work-box. You would then be able to move the dirt to each jar in the case the gaps for your antibacterial cleanser washed hands. Wash the spoon used to move the dirt after each jar, so that polluters don't go from jar to jar.

B: Close the rye jar cover firmly. Wash the outside of the cover with antibacterial cleanser and Lysol. Do likewise with a dirt jar. Remove the part from the jars, however, leaves the tops on. Turn the dirt jar holding the cover on and place it on the rye jar. Covers ought to look against one another. Shower the territory. Then, cautiously slide the two covers away, letting some dirt fall through into the rye jar. Be mindful so as not to give a lot of fall access. It may is not necessary to slide the covers right off to get them to fall in. Slide the tops back on when the rye is secured.

10) Cut the plastic board to fit over the cooler. Wash and Lysol the board. Restore each jar with tin foil up to the highest point of the dirt. Expel every cover simultaneously place a zip lock pack over the opening. The opening of the sack should cover the opening of the jar. Along these lines, air can get in it is as yet secured. The jars can be handily circulated air by sliding here and there over the container gradually. Jars can be watered by staying spout of the shower bottle in under the edge of the pack. Along with these lines, the jar is rarely revealed. Spot all the shrouded jars in the cooler and keep the cooler in a perfect area. I have discovered that these packs sterilized soil are the way to sterility. The plastic top certainly not keep that much out. Each time you open it, a wide range of residue glides. I have had **NO** sterility issues on jars that were stowed and utilized soil. Never take the pack off anything else than sufficiently only to stick the spout of the splash jar in there. Circulate air through the jars as depicted dust gradually doesn't get sucked into the jar.

11) At this point, the jars should be splashed with pure water every day. Try not to make it to be excessively wet. However, following a week or so, you should see mycelium start to cluster together at the edges of the jar. They should consider more in the following week as they develop into the dirt. Keep clouded with a neat shower of water. If they develop too thick, you should see them somewhat heavier to wreck. Try not to splash the dirt too, as you would welcome different molds. In about fourteen days from packaging, the mycelium from the rice cake will have become through the dirt and may begin to get through the highest point.

If this happens, shower the dirt somewhat more to wreck them. The cooler ought to get around **12-13** hours of light a day through the top. The surrounding room light is fair. Keep it out of the immediate daylight, so it doesn't get excessively hot. Keep watching for any form and be set up to expel them from the cooler promptly. The shape can be hard to spot, so be careful.

Psilocybin Mushrooms

The most widely recognized molds to search for are a green form and a yellowish sludge. If a jar is polluted, check the jars, it was sitting close. They might be defiled well. Consequently, it is a smart thought to space the jars as far separated conceivable in the cooler. Try not to attempt to rescue sullied jars; it won't. If contamination is discovered, wash off all the exterior containers with bacterial cleanser. Change the tin foil and shower the cooler and cover Lysol before supplanting the great jars.

12) The initial flush of mushrooms ought to show up in around 2-3 weeks after packaging. The jar will keep on delivering mushrooms for 40-60 days. Pinheads start as minuscule white spots and develop into what resembles scaled mushrooms with heads and thick stalks in a day or two. Shrooms develop from pinheads to full mushrooms in about seven days. At the point when the edge of the top isolates from the tail, it is prepared to harvest. Use tweezers to snatch the base of the tail and squirt it out. It is likewise a smart thought to fill the gap that is left with new packaging soil. It will make the jar's natural product longer. Pinheads may frame underneath the dirt close to the glass and never get through to the surface. These can be evacuated and the gap is loaded up with packaging soil. After the principal flush mushrooms have developed and the square of the rye has pulled away from the sides of the jar. **O&O** exhort uncovering the packaging soil and reworking the entire thing. There are a few changes which you will see in the jar as it develops. You ought to anticipate that these things should occur if the shrooms are solid. These are significant changes arranged by event alongside some other arbitrary recommendations:

- After packaging, ropy sprinters will show up close to the edge of the glass in the dirt. These will obscure in shading into yellowish earth-colored. Try not to confuse this with a disease. I think it is the shade of the supplements which the sprinters are conveying all through the jar. Sprinters in the rye will remain white.
- Some pinheads will become further in the jar regardless of the tin foil wrap. It appears to have no clue about what direction is up. Try not to stress over it. If you attempt to expel, you'll most likely pollute the jar. They'll quit growing and return to customary mycelium.
- When the mushrooms initially develop, they will give off an impression of being thick. At the point when they are to open, the tail close to the top will contract and the top will get somewhat bulbous (quick and bulbous). It is typical. Your mushrooms are dying. They should get more slender.
- I have seen an inconspicuous contrast in the manner mushrooms react to light. They appear to become taller in obscurity and thicker in the light. May simply be fantasizing, however.
- There will be a small line of obscuring where the top meets the cloak not long before the shroud tears open. I think this is because of wounding. Shrooms turn blue when wounded.
- Most of the mushrooms develop close to the edges of the jar and some even on the jar. If clusters of white mycelium

develop on the jar, disregard them. The absolute best mushrooms originate from them.
- If the mycelium truly congests the highest point of the dirt, I have thought that it was helpful to include another inch of packaging soil hardly. It may be redundant, but rather it works for me.

Chapter Fifteen

Tek for Magic Truffles - Truffle Tek

Necessities for Truffle Tek:

- Quart bricklayer jars with covers
- Filter material - just engineered channel plates or poly-fil stuffing
- Silicone
- A substrate, either Rye grain (Rye berries), Whole Oats (absolutely entire) or even Wheat grain (wheat berries)
- Pressure cooker
- Species spores/culture
- Syringes w/needles or still-air box or clean laminar wind stream bureau, contingent upon your strategy for inoculation

Setting Up the Grains:

It is the center of undertaking the main mushroom, included food. As the mycelium spreads through it, the supplements in the grains and water they contain truly become the mycelium and sclerotia. Therefore, the significance of having grains wealthy in sustenance, the correct level of hydration and size/structure that makes it simple to separate the sclerotia from the grain.

Rye grain (rye berries) performs best in all offices.

Ultimate **oats** are not a long way behind.

Wheat berries work nearly as magnificently for similar reasons and are significantly progressively cheap.

While **corn** is a poor grain for all other mycology tries, it is a fine sclerotia grain, however increasingly costly and calls for progressively cautious readiness.

Wild feathered creature seed (WBS) is adequate if very much flushed, deliberately hydrated as the numerous seeds are of various sizes and sunflower seeds are evacuated.

There are numerous approaches to stir up one's grains, yet the **methodology** introduced here is straightforward and genuinely hard to fail.

You can work on heating your grains with a lot of additional water in a big pot with a hard moving bubble until they are full and soggy and burst decently between your fingers. Yet, a large portion of the structures is unmarked, you will, in any case, have various. They pop, which is fine. Don't stress except if a considerable bit of them have popped. Heating the grains hard makes it simple to get ready as you don't need to mix them - the agitating water will do this without anyone's help.

If you need to have as hardly any popped grains as could be expected under the circumstances (I don't discover it makes a difference much. However, it's as yet pleasant), at that point adopt the more involved strategy of naturally blending with a light bubble (Simmer). You do need to mix regularly to ensure you don't over-hydrate and pop the base grains. If you don't mix much of the time, you could wind up with **MORE** burst grains like this. Recall that it will be a more drawn out procedure since the grains will hydrate and grow all the more gradually.

Focusing on the most significant water content without going over is perfect. Having too little water brings about hard to colonize grains and a poor sclerotia yield. Burden grains in a sifter, **"shake"** downwards over the tub or sink to help trickle channel and set in the sink to deplete more. Channel the grains well, hurling the sifter and back up to toss any extra water out the base of the sifter. If they aren't excessively wet, which you'll figure out how to remember, they're prepared to stack up and sterilize when they're finished steaming. Odds are, the length of your grains is full, all-around hydrated and not standing any kind of test outwardly, they'll do incredible in the pressure cooker.

Did you decide to explore different avenues, including plant food? If you did, an all-inclusive adaptation of this **'aging'** period is essential with heat treatment in the middle, as microbes fit for taking up nitrogen and phosphorous into different structures that permit mycelia the capacity to use it progressively.

Setting Up the Jars:

It is basic. They should have sterile air filtration so both jars can be shut during pressurized sterilization of jars and substance. Thus, that excess **CO_2** inhaled out by the mycelium can be pushed out and traded with the air without permitting any attacking spores or microscopic organisms. Manufactured channel circles (SFDs) are ideal for this all around. Poly-fil designed stuffing can be utilized with full achievement when pressed firmly.

Drill an opening at any place in the tops. When utilizing **SFDs,** "1/4-3/8" are ideal with nickel or quarter-sized circles. Openings near "3/8" are significant for utilizing poly-fil. If you will use syringes for your inoculation methods, a little second gap is also required to immunize through.

Follow and cut the circles. Put an even measure of silicone around and directly off the edge of the cover's gap and spot the plate focused over it. Pushing don the edges of the circle to protect it with the silicone, regularly presses some of it into the space of the gap. Ensure that silicone doesn't hinder the ventilation - cautiously wipe it away with the collapsed corner of a paper towel or preferences.

If you penetrated the subsequent little opening for immunizing by the use of a syringe, at that point, include a pleasant smear of silicone over and around that gap on the two sides. The needle consistently slips in and out and doesn't permit entry of sullying except if it's on the needle or the silicone itself, which you ought to take care of with care before immunizing. Your tops are finished. Set them over their jars for the silicone to dry. At the point when prepared, load the jars with the grain. For your first development, I recommend you load them to hardly around the **600mL** imprint.

Cleaning Jars and Grain:

It is most likely the selective part, yet you have to ensure you pressure-cook appropriately. Try not to let the jars sit on the base of the cooker. Save rings make almost flawless struts. Include water in the last, 1/4-3/4 inches, as to not come up to the highest point of your spacer. Run your oven on **"high"** to get this show on the road.

If you are immunizing with a syringe or in a glove box, you may think that it's supportive guaranteeing a sterile injection to cover the highest points of the jars with a square of foil. So contaminants don't land around the highest points of them. Thus dampness on/in the channels from the cooker doesn't wick in pollutions previously or during inoculation. Burden the jars in and put the top on the pressure cooker. Proceed with the oven running maxing out until the cooker is up to 15 PSI pressure, which is full pressure on most big cookers.

Set a clock for 100-120 minutes once the pressure is reached. Lessen the temperature as it is necessary to keep **15-16 PSI** marginally. As the jars and their substance heat up over the cooking run, periodic decreases in oven temperature will presumably be called.

After 15PSI has been supported for 100-120 minutes, turn off your oven. Permit pressure to drop normally - don't let loose a little. Allow the temperature to drop normally too. It will take a few hours, likely as much as 8. The first activity with the jars after sterilization is to shake the grains around, equitably circulating any grains with more dampness that may have been against the sides or bottoms. When cooled and dampness entirely settled, the grains will look as above. They're prepared to immunize.

Inoculation

The critical advance. Here the moment of truth will represent your undertakings. If a solitary form spore or bacterium makes it onto sterile grains, it will take over every last bit of it, that jar demolished. Significantly, the mushroom spores or societies are the main things that make it to the substrate. It can be made simple and guaranteed by a couple of specific methods.

In mycology, there is about an innumerable plenty of sources from which and approaches to vaccinating. Any of them will fill in insofar as you're ready to succeed looking after sterility, uncovering neither the inoculant nor the substrate to tainting anytime. Obviously, certain methodologies have different focal points and downsides. This talk will lay out the nuts and bolts of those that best suit the finishes of these tasks, barring those lab-style and preferring those including just the straightforward procedures as of now close by.

Spore Syringes are the least demanding and most regular way societies are begun for truffle devotees. Clearwater contains thousands and thousands of spores that sprout upon food and start extending. The distinct mycelia must be offered the chance to mate before the undertaking of colonizing the grains to deliver truly skilled hereditary strains for our motivation. Hence, you should shake your grains BEFORE inoculation and NOT shake them once, including the spores! Level-off the grains to vaccinate. Accomplish all spotless work in a perfect spot without any drafts. Outside, just after a downpour is a perfect cool.

Shake your syringe well. With a customary lighter, burn lighter or liquor burner, completely fire sterilize the needle. Promptly expel the foil on the jars and stick the needle through the silicone inoculation port. Each jar just needs a modest quantity of spore arrangement; .5-1mL is bounty to guarantee satisfactory hereditary assortment. Spurt the arrangement of the side, sufficiently only to hurry to the base, in one to four places on the jar. Once more, don't shake the grain to convey spores. It is best that they remain close by.

The settlement in the jar on the left is around the prime size to be shaken and uniformly dispersed all through the whole substrate. It might require a few firms smacks against the palm to separate a combined development spot. The jar on the correct needs more time before a shake is worthwhile. The mycelium will appear to vanish. Put something aside for a light white dimness on the outside of grains that were colonized. Try not to stress! The mycelium on/in those will recoup and rapidly starts new settlements to cover the rest of the grains.

At the point when the substance of the jar is barely colonized with solid white mycelium, the jar is finished. Now it tends to be determined to a rack to be calmly once in a while seen through sclerotia arrangement or utilized as a hotspot for handfuls progressively perfect inoculations of different jars.

There are TWO essential **approaches** to utilize them all things considered, arranged by tenderfoot availability:

1. *Water Suspension:*

It is an excellent technique for home cultivators without a despite everything air box or clean laminar wind current bureau, utilizing the silicone inoculation port on the jar covers, the everything air box can be prescribed because of client blunder or an undermined syringe. It is basic and profoundly clean and dependable as long as the source jar contains NO contaminants evident or covered up and the grains are not near over-hydration, which isn't a concern if you did the grain prep of this Tek.

This procedure will appear to be dull except if you get a quality 30mL or 60mL syringe!

Fill a sifted and ported jar with 200-400mL of water and spread the top with foil and wrap an unfilled needle-tipped quality syringe with foil. Pressure sterilize them for 40 minutes at 15 PSI. Permit pressure and temperature to fall normally until room temperature. Try not to utilize the water when hot!

In a spotless spot, rapidly expel the foil from both and plunge the needle through the inoculation port at that point to attract up water to fill the syringe. Splash liquor on the inoculation port of your colonized ace jar, enough where it covers the port (instead of just many standing beads), however insufficient where you'll take drops through the port.

Fire sterilize the needle and promptly dive into the ace jar's port at that point, infuse the sterile water into the jar. Include at any rate 40mL along these lines, up to 100 for a solid extraction. Now, without turning the jar on its side, shake it to relax the grains and wash mycelium off. If the jar has been colonized altogether for more than a couple of days, this may require some firm strikes. If possibly more than seven days, it might be very hard for separation. However, you will prevail as long as the way of life has not started truffle development. Try not to separate the mycelium before including water!

The grains will ingest a decent piece of water, so if there isn't a lot of free water after the shake, at that point, proceed with fire disinfecting the needle between each wound through the port and moving sterile water to the jar. Tilt the jar towards the inoculation port with the goal that the pool of mycelium-loaded water comes in reach of the needle at that point, suck it into the syringe.

There you go! That water suspension of live mycelium can be utilized straightforwardly to vaccinate new jars or spurted over into the first jar of sterile water. Weaken it to some degree and amplify its solid potential.

2-10mL can be immunized in each jar (don't infuse excessively if the new grains are near being too wet as of now) and the jars were shaken altogether to disseminate the mycelium.

It is better and extraordinarily ideal than sugar-took care of "fluid societies." Any fluid inoculation will take 2-6 days to recoup and start building up itself in its new food. However, it will colonize the grain rapidly! It permits brisk, solid and even arrangement of truffles!

2. Grain-to-Grain Transfer:

It is an amazingly successful approach to extend societies and is presumably utilized the most. It requires a spotless air condition, for example, a despite everything air box or clean laminar wind current bureau.

With your sterile exchange workspace and yourself prepared, just separate the grain in the colonized ace jar and shower/wipe it all together with liquor, in the still air box expel all rings from all jars included and afterward the cover of the ace jar. Hold the ace jar in one hand and individually evacuate the top of another sterile jar of grain rapidly and spill out a touch of the ace jar's colonized grain. Quickly supplant the new jar's cover and move to the following one! Give a valiant effort to isolate the ace grains among the new containers equally.

The significant part is finished! Presently simply fix the rings back on the covers and thoroughly shake to disseminate the vaccinated grains among the new ones equitably.

They are utilizing a decent measure of inoculant grains, which guarantees that the substrate will be thoroughly colonized shortly and that sclerotia will frame unequivocally and equally.

Keep your jars in room temperature 68-75F. They need no "incubators." However, they will go significantly more gradually beneath 65F. Surrounding lighting can help speed truffle improvement, so maybe turn the jars once in a while to apply this upgrade uniformly. When truffles are uniformly conforming to the jars, there isn't generally any profit by light.

Help yourself out. Try not to open that jar. Hold up until one week from now. One more week as of now, eh? Indeed, do feel free to hold up one more. I know, I know, however, trust me. Hold up one more week.

Harvest

A feel burnt out on some sort is crucial to separate the 3-month-solidified cakes of sclerotia-loaded rye. Marginally emptied bicycle tires are great. In case you're reasonably cautious and utilize just these, you'll abstain from having a glass jar break in your grasp. Except if spore change shows the infrequent low-maker in test ace culture jars, Ps. Galindoi and Ps. Mexicana will create more than 100 grams of new sclerotia, ordinarily as much as 125g, up to an exceptional max ~150g.

New sclerotia harvested now are regularly found the in middle value of Twice (x2) the strength of run of the mill new Ps. Cubensis mushroom. The mycelia do have double the normal psilocybin digestion by and large, as the mushrooms it shapes regular twofold run of the mill Cubensis likewise.

With the truffles, it is to some extent because of the lower water content per body mass proportion of sclerotia than mushrooms, as the sclerotia mycelium has not gone into the fruiting stage. Mushrooms of the two species are 9/tenth water, which means the magnificent accommodation of almost moving a decimal spot over from new to dry loads. Truffles anyway are ~2/3 water, so they contain hardly over triple the got dried out contagious mass than mushrooms.

It implies that got dried out sclerotia are normal 2/3 the strength of run of the mill dried Cubensis. Even though organic digestion consistently shifts, the accompanying model weight counterparts for regular examples of every species appear to be incredible thumb rules:

- 140g new Galindoi sclerotia = 283g new Cubensis mushroom = 47g dried Galindoi sclerotia = 28.3g dried Cubensis mushroom

- 5g new sclerotia = 10g new Cubensis = 1.66 dried sclerotia = 1g dried Cubensis

For the great act of standard micro-dosing, .5-1g new sclerotia are great, start low.

The best part is that you don't need to harvest them by then. They can stay in the jars for quite a while still. With a perfect grain readiness, it is likely they can go at any rate a half year, conceivably even a year relying upon how well the channel holds dampness. The mycelium keeps on eating, keeps on breathing and keeps on processing. Those metabolites are bitten by a bit consistently taken care of into the sclerotia bodies after some time. Although this domain is less investigated by other than some truffle organizations of the world, contemplations recommend a quality increment to around x3 that of new Cubensis.

Chapter Sixteen

Cultivation Equipment and Supplies

The cultivation of Psilocybin Mushrooms is a tricky process. You have to be prepared for the interference by other organisms. But if you have thoroughly followed the **S.O.P.s** for cultivation, you should be confident. Different equipment and accessories are required for the development of mushrooms at the laboratory level or broad levels. You need all the supplies, even if you are going to grow mushrooms in your home yard.

Mushroom cultivation requires both simple and specialized equipment. There are mushroom supply and mushroom cultivation supply houses from where you can get the things required for the cultivation of mushrooms. Some materials and apparatus you need are available in other shops like a pet shop, pharmacy, etc. You can also get the required equipment from hardware stores and kitchen supply stores. The stores from where you can get the equipment needed for the cultivation of **Psilocybin mushrooms** include:

- Pharmacy
- Garden center
- Mushroom cultivation supplier
- Hardware store
- Pet Supply
- Home Brewing
- Kitchen supply
- Medical supply
- Scientific supply house
- Supermarket

You will need a list of different things from the supplies mentioned above and stores. After collecting all the equipment and supplies, you will be able to start mushroom cultivation. Let's discuss the major equipment required for the growth of mushrooms.

Petri dishes:

Petri dishes are one of the main utensils required for the cultivation of mushrooms. The mushroom cultures are prepared in Petri dishes and they are an important part of the microbiology laboratory. A Petri dish comes under the category of glassware. It is a glass or plastic object, sees through and shallow and has a loose-fitting cover. Petri dishes of different sizes and qualities are available in the market. Reusable Petri dishes are also available, but they are expensive. So if you have a high budget, then you can go for reusable Petri dishes. The pre-sterilized Petri dishes, which are disposable and made up of polystyrene, are available in the form of sets of 20 to 25 dishes. These dishes are not expensive and they are affordable and economical. But, at the same time, they are not economical because they are not reusable and are not environment friendly. Although the Petri dishes are sterilized, we should clean them again, by using a microwave oven and hydrogen peroxide. The steps of **sterilizing** the Petri dishes are:

- Use dishwashing detergent and clean all the Petri plates but take care while washing.
- Take a little quantity of peroxide, about **3%** and pour it in each dish and swirl the dish to reach the chemical to all sides of the Petri dish. Do the same with its cover.
- After washing this way, open the microwave and put that stack of dishes in it with medium power, put them until the peroxide dries.
- After sterilization, pack the plates in plastic bags or use them immediately.

It is the most useful and effective method of sterilization, especially when working with agar. Note a tip that you can use the hydrogen peroxide in your cultures and it will reduce the chances of contamination. It is not recommended to use hydrogen peroxide when spores are germinating. You can use the Petri plates with a **50mm** diameter for preparing mushroom cultures. But if you have a lack of supply of Petri dishes and they are unavailable, you can use some other things like Jelly Jars (of 4 ounces) or glass containers. These supplies have a positive point that they are reusable. But they cover more space and are not see-through like Petri dishes.

Pressure Cooker:

You may ponder that why we use a pressure cooker in the cultivation of mushrooms by disinfected culture methods. If you know about the orderly uses and advantages of a weight cooker, then you can get the point right. Weight cookers will be utilized ordinarily in cleaned culture methods of mushroom development. Continuously use the excellent weight cooker else it might cause harm. The weight cookers utilized in labs are somewhat not the same as the cookers utilized as kitchen products. Utilize a bigger cooker for lab use, yet if the large isn't accessible, you can utilize the medium measured cooker. The size additionally relies upon your work, what amount of containers you are going to place in that cooker? Purchase the Cooker as indicated by it. We like to utilize the weight cooker fabricated by some American Brands like **Wisconsin Aluminum Foundry**. They give the best and dependable weight cookers. You can pick the weight cookers relying on your requirements. Various kinds of weight cookers are accessible for the most part.

Some have a valve, the steam discharge valve.

Some have a rocker, comprised of metal. It begins to vent out the steam on a specific weight edge.

The second type can be risky because fluids begin bubbling in such weight cookers and may demolish the media.

The stopcock type is called the **sterilizer**.

The rocker ones which are additionally called **canners**.

Bricklayer Jars:

You can utilize Mason containers for the development of societies. The Ball-style bricklayer containers comprised of quartz are effectively accessible. Limited mouth bricklayer containers are best for grain production. Continuously check the artisan containers for splits and so forth before use. For the most part, these containers are sturdy. However, they may get breaks some of the time, so check this before developing your societies.

Media Flasks:

Media Flasks are basics of a lab like Petri dishes. You can develop your mushroom societies in media flagons if the way of life is in fluid structure. During cleansing, media cups can, without much of a stretch, keep the liquid media. Moderately limited mouth bottles are useful for this reason. You can utilize squeezed apple bottles with tops, which are perfect to use as media jars.

Covers of Mason Jars:

Bricklayer container covers are significant and they ought to be cleaned before use. The plastic top, which ought to be heat safe, is ideal to use as Mason container top. The covers ought to be autoclavable. Make a one crawled gap in the top and make it fit with the channeling circle and let your way of life-breath without any problem.

Patch Sacks or Produce Packs:

The channel sacks or bring forth packs are adaptable and heat-safe plastic packs that can undoubtedly hold production in large amounts. These sacks are autoclavable and have a channel on a side for the trades of gases. You can place the substrate in it and afterward sanitize and brood. At that point, seal it. These sacks are adaptable. Accordingly, they are utilized for enormous amounts of bringing forth. We can analyze the substance of these sacks effectively for contaminants and so forth.

These sacks, when warmed, for the most part, lose their versatility. Accordingly, they are beneficial for one-time use. The packs of good quality, which stay fit as a fiddle significantly after one-time disinfection, can be utilized twice. A few ranchers and mushroom cultivators use stove packs to place bring forth in it. Be that as it may, these broiler sacks don't have space for gas trade. They are additionally meager and you can't redistribute the substance in stove sacks. So it's smarter to utilize explicit bring forth or channel fix packs. If you are getting ready to generate in large amounts, you can use the big bricklayer containers.

Filter Discs:

If we talk about the equipment used in the cultivation of mushrooms, filter discs are also significant. They are used to cover the space between the lid and the mouth of the jar. These discs are responsible for allowing the gas exchange. Filter discs are mostly made up of synthetic fiber, which is a heat resistant material. They can also be cut to fit the jar. **Tyvek** is the material which can be used as the economical replacement of filter discs. Tyvek is thinner than filter discs. So, when we put it in between the lid and mouth of the jar, it is cut wider and hang on the edges of the jar.

Alcohol Lamp:

An alcohol lamp is also a crucial part of a laboratory. An alcohol lamp is made up of glass with a metal collar. Alcohol is filled in the lamp and a thread is protruding out from the mouth of the lamp. This thread is ignited and then it provides a bright flame. As the replacement of an alcohol lamp, you can also use a mini torch. Mini torch contains **butane** and it is available in all-electric stores. An adequate mini torch has a good solid base.

Funnels:

You should have both metal and plastic funnels, but they all should be narrow-necked. These funnels are used to pour liquids and powders handily in the jars etc.

Balance:

Digital balance, mechanical or electric balances, are all necessary before starting mushroom cultivation. Always use the balance, which shows the accuracy of about 0.5g and can weigh up to 250g. Use a balance with a broad pan to adjust all the things.

Scalpel:

A Scalpel is used to cut and transfer the cultures. Always use disposable scalpels to avoid contamination. A disposable scalpel with a blade of size ten is promising for your cultures. An aluminum X-Acto style knife can be used as the replacement of the scalpel. But if you have scalpels, it is best.

Inoculation Loop:

You are already familiar with the wire loop, which is used to transfer spores or mycelium to agar plates to grow the cultures.

It is made up of metal or wood with a small metal loop at the end. You can get it from scientific stores or brew making stores. You don't need a wire loop if you are using the cardboard disc spore germination.

Graduated Cylinders, measuring cups:

Graduated cylinders are necessary for measuring the exact amount of media. Mostly the cylinders of 1 liter, 100 milliliters and 10 milliliters are used, which can cover all the quantities. Measuring cups and spoons are, also important for measurements. You cannot grow a perfect culture if your measurements are not accurate and you are not using the exact quantities. All these measuring equipment should be sterilized first because they may contaminate the cultures.

Syringes:

Syringes are used for the inoculation of cultures. These common syringes are available at all surgical or veterinary supply stores. Also, sterilize the syringes before use.

Summary: Above mentioned pieces of equipment are required for the cultivation of mushrooms and preparation of cultures of psilocybin mushrooms. You can get all this equipment and apparatus from any scientific store and some parts are also available at grocery stores. Supplies are different, which we will discuss below. Equipment and supplies both are necessary for starting the culture preparation and cultivation of Psilocybin mushrooms.

Supplies:

Hydrogen Peroxide (3%):

Hydrogen peroxide is an antiseptic and it is used in the cultures to prevent any contamination. You can get hydrogen peroxide from grocery stores or pharmacies. The concentration of Hydrogen peroxide may differ from store to store, so always check and confirm concentration (3%) before using it. This concentration is not harmful to humans, so it is not necessary to take special measurements while dealing with hydrogen peroxide. At stores, 8 to 35% of concentrated solutions are also available. So you should be careful about concentration. But even if you are using 3% concentration, wear gloves and prevent your clothes. Concentrations more than 3% can cause burns and it is flammable, so avoid such engagements. If you are diluting the hydrogen peroxide, take specific measurements while dealing with it.

Isopropyl Alcohol:

What will you use as a fuel for alcohol lamps? The answer is Isopropyl alcohol. This substance is not only used as a fuel but also for disinfecting your hands and surfaces. You can get

Isopropyl alcohol from pharmacies and grocery stores. You can use any specialties available in stores in concentrations of 70% and 91%. While using isopropyl alcohol, you must take special care because it is flammable. Keep an alcohol lamp away from isopropyl alcohol.

Bleach:

To clean the surfaces and tools thoroughly, you will need a bleach. Easily available laundry bleach is enough for this purpose. There is no need to use the bleach added with detergents as it may damage your skin and tools. For cleaning, you can dilute it and for disinfection, you can use a 100% spray of bleach.

Parafilm:

Parafilm is used to seal the Petri dishes and is made up of paraffin. It is an elastic film which allows the gas exchange but keeps the contaminants away. You can also get it from garden supplies as grafting tape. You can use polyethylene cling film as a substitute for paraffin film but avoid to use Glad Wrap because it is not gas permeable.

Surgical gloves:

Surgical gloves are also available easily at grocery stores and pharmacies. These gloves prevent your cultures from contamination through your hands and also prevent your hands from damage and burns etc. Always sterilize the gloves before use and wash your hands before and after the use of gloves. Also, clean your gloves with a towel paper soaked in alcohol.

Substrates for Psilocybin Mushroom Cultivation:

Whole grains:

Mostly the cultivation of mushrooms is started by the production of spawn. If you are producing spawn, the best substrate for spawn production is "whole grains." Spawn is a mixture of mycelium and substrate. Grain plays the role of the best substrate because every grain acts like a capsule full of minerals, nutrients and water. It can be simply colonized by fungi, like mushrooms. The fibrous husk of grain protects the fungi from other organisms and prevents contamination. When colonized, grains are then separated from each other. After colonization, we use the grain spawn for vaccination and grain acts as a reserve of nutrients.

You can use any grain, but it is recommended to use the grains of white winter wheat because, according to experiments, white winter wheat has acted as the best substrate for the growth of psilocybin mushrooms. Another advantage of using wheat grains is that they are free from contaminations from other organisms. Other grains that you can use include rye, corn, millet, etc.

Dried, malt extract:

Malt has been used as an adequate substrate for growing cultures for years. The grains are malted and their starch is converted into partial sugars. The primary nutrient source of agar media is malt extract. You can get malt extract easily from brewing suppliers. If the malt is darker, it means that sugars are caramelized. If the sugars are caramelized, fungi don't grow on such media. So don't use darker malts.

Yeast extract:

Yeast is also a nutritional supplement, which is added to the media. It is a good source of vitamins, minerals and proteins. You can easily find the Brewer's yeast at any food supply store or the grocery store.

Lime or Calcium carbonate:

Calcium carbonate is known to us with different names like lime, hydrated lime, oyster shell flour, limestone flour and chalk. It has different uses like it is used for the buffering and maintaining of pH of soils and substrates, keeps the contaminants away and acts as a source of Calcium for growing fungus. Fungi like to grow in slightly basic media of pH eight, but bacteria and most of the other microorganisms do not grow at this pH.

CHAPTER SEVENTEEN

Micro-Dosing Psilocybin Mushrooms

Micro-dosing is the demonstration of expending sub-perceptual (unnoticeable) measures of a hallucinogenic substance. Numerous people who have coordinated micro-dosing psilocybin mushrooms into their week after week schedule report more significant levels of inventiveness, more vitality, expanded center and improved social abilities, just as reduced stress, anxiety and even depression. A few aficionados likewise report that micro-dosing psilocybin has helped them uplift their profound mindfulness and improve their faculties.

Hallucinogenic specialists have likewise found that psilocybin can affect disposition issues and anxiety. Mushrooms (and LSD) have additionally been found to have practically identical or better outcomes in treating bunch cerebral pains than most traditional prescriptions — numerous individuals have encountered expanded times of reduction in the wake of treating their migraines with hallucinogenic substances. With all these empowering results on full doses of psilocybin, there's motivation to accept that micro-dosing could realize positive life changes likewise.

How Do You Micro-Dose With Psilocybin Mushrooms?

Micro-dosing with psilocybin mushrooms is a genuinely direct procedure. You have to set up your micro-doses, expend the micro-dose at the appropriate time and follow a month-long convention to guarantee you experience enduring advantages. We layout every last one of these components in more detail underneath.

Setting up Your Micro-dose:

Getting ready psilocybin mushrooms for micro-dosing includes a more significant number of steps than micro-dosing with **LSD**, yet it is clear. The trickiest part is assessing how much psilocybin is in a specific mushroom. Various strains of mushrooms won't just have different amounts of psilocybin; however, new and dried psilocybin mushrooms do, as well. Indeed, even varied pieces of the mushroom contain marginally various sums.

In any case, we suggest drying a cluster of mushrooms, pounding them into a powder and apportioning around 0.1g of as your starter microdose. From that point, you can modify the sum in like manner. In the end, when you take a dose that causes you to feel a few changes (most remarkably laziness, the primary impact that enters a psilocybin trip), move it back to solely under that sum. That is your mushroom micro-dose sweet spot. You can utilize any sort of psilocybin mushrooms for micro-dosing. The most famous ones are Psilocybe Cubensis, Psilocybe Semilanceata, Psilocybe Azurecens, Psilocybe Cyanescens and Panaeolus. Simply know how much psilocybin content your strain has (for instance, the last two are wealthy in psilocybin) and modify your microdose likewise.

Step by step instructions to Take Your Micro-dose:

There are a few different ways to take your micro-dose. The most reasonable one is to quantify your doses into void cases. It will guarantee even circulation and veil the taste. Another choice is to set up a psilocybin tea by dissolving your ideal dose of powder in high temp water and any event, including some nectar. Be that as it may, don't hesitate to test and blend the powder into any beverage you take to begin your day.

What Micro-dosing Schedule Should I Follow?

Specialists around there recommend distinctive micro-dosing regimens. It is suggested that taking a micro-dose once at regular **intervals:** Take a micro-dose on Day 1. At that point, don't take a micro-dose on Day 2 or Day 3. On Day 4, take another micro-dose. Proceed with this procedure for a little while.

For many people, the morning is the best time because the valuable impacts will last for the day without meddling with rest. It's likewise useful to take everyday notes in a diary to watch the effects all through this procedure and modify as needs are—or simply notice the positive changes.

It's additionally critical to follow your typical daily schedule while micro-dosing. The design is to improve your everyday presence by coordinating micro-dose into your daily practice, so don't change what you typically do. Be that as it may, when you have a go at micro-dosing just because, take a vacation day from work and social duties. It will allow you to see any unordinary impacts before micro-dosing in a progressively open circumstance.

While it might appear as though you would just feel the impacts of the micro-dose when you take it, attempt to watch the effect on the two days between doses, as well. Numerous individuals see expanded sentiments of stream, inventiveness and vitality the day after they micro-dose, notwithstanding the day of micro-dosing.

Psilocybin and lion's mane both can make new neurons and neural pathways and to fix existing neurological harm. Out and out, this one of a kind blend of mixes can be fused into different treatments with such combinations giving one of a kind focal points to therapeutically huge headways in fixing neurons, evacuating amyloid plaques, improving psychological well-being, insight, readiness and generally enhancing the nature of cognizance.

Micro-dosing consistently isn't suggested. Since your body creates resistance to **psilocybin**, you may see unavoidable losses following a couple of days if they are taken each day. It is the reason **Fadiman** proposes leaving a few days between each dose. Also, the way that constructive outcomes can now and again be felt many days after a micro-dose is a valid justification to scatter your treatments.

Another drawback to micro-dosing consistently is normalizing a potent substance. You can contrast it with the utilization of espresso for efficiency purposes. At the point when you drink espresso consistently, after some time, you have to expand the dose to get a similar impact. Inside several months, one cup transforms into two, three or four cups. It is ideal to use micro-dosing as a periodic bit of leeway, as opposed to a steady go-to like espresso.

Advantages of Micro-dosing:

There are numerous advantages to micro-dosing. In any case, at the danger of distorting, the vast majority micro-dose for two principal **reasons:**

1. To decrease the recurrence and power of unwanted states brought about by different types of psychological maladjustment, including:

- Depression
- Anxiety
- ADD/ADHD
- Mood issue
- PTSD
- Addiction

2. To expand the recurrence and power of attractive states/results, including:

- Creativity
- Energy
- Flow states
- Productivity/center
- Improved connections/expanded sympathy
- Athletic coordination
- Leadership advancement

Alluring States:

Numerous individuals micro-dose for self-awareness or motivations behind self-streamlining. Reports propose that the training can improve imagination, efficiency and vitality. Innumerable individuals likewise micro-dose to assist them with tackling business-related issues, make new ideas or just to decrease hesitation. Micro-dosing can also help you by improving your social communication aptitudes, athletic execution and otherworldly mindfulness.

Dangers:

By a long shot, the most dangerous thing about micro-dosing is the constitution. It's urgent to check your nearby laws before micro-dosing, as the punishments for the ownership of psilocybin mushrooms are as yet brutal in numerous nations. It's as yet conceivable to micro-dose psilocybin mushrooms lawfully and we never support illicit activities. Aside from the legitimate dangers, micro-dosing has demonstrated to be a safe, non-compromising prologue to the advantages of hallucinogenic. Psilocybin has a long reputation for safe use. Consolidate that little dose sums and micro-dosing gives off an impression of being sheltered.

Leadership and Micro-dosing:

Change and advancement are occurring quicker than at any other time nowadays and pioneers need to adjust rapidly. Incredible pioneers must think of imaginative answers for surprising issues and issues making potential mishaps advantageous for them. Remaining at the bleeding edge of any field requires acing innovation and being available to novel methods of achieving errands, considering both short and long haul needs. Moreover, present-day leadership is getting less various leveled, prevailing and forceful (prototype "manly" attributes) and progressively requires the capacity to develop space to permit the best individuals to step in and contribute their most noteworthy endowments (original "female" qualities).

Micro-dosing assists with quickening this formative procedure for the up and coming age of pioneers by encouraging upgraded inventiveness, mental adaptability and legitimate self-reflection. Micro-dosing likewise develops self-assurance, which empowers you to be more in contact with your feelings and helps you impart better

www.ingramcontent.com/pod-product-compliance
Lightning Source LLC
Chambersburg PA
CBHW060839220526
45466CB00003B/1166